THE DEEP SEA

THE DEEP SEA

JOSEPH WALLACE

GALLERY BOOKS
An Imprint of W. H. Smith Publishers Inc.
112 Madison Avenue
New York City 10016

A FRIEDMAN GROUP BOOK

Published by
GALLERY BOOKS
An imprint of W. H. Smith Publishers, Inc.
112 Madison Avenue
New York, New York 10016

ISBN 0-8317-2177-4

THE DEEP SEA
was prepared and produced by
Michael Friedman Publishing Group, Inc.
15 West 26th Street
New York, New York 10010

Editors: Sharon L. Squibb/Emily Loose
Art Director: Mary Moriarty
Designer: Robert W. Kosturko
Photo Editor: Philip Hawthorne
Production Manager: Karen L. Greenberg

Typeset by BPE Graphics, Inc.
Color separations by South Seas Graphic Arts Company Ltd.
Printed and bound in Hong Kong by Leefung Asco Printers Ltd.

DEDICATION

For my brothers, Jonathon and Richard,
and for Sharon, who sees everything first.

C O N T E N T S

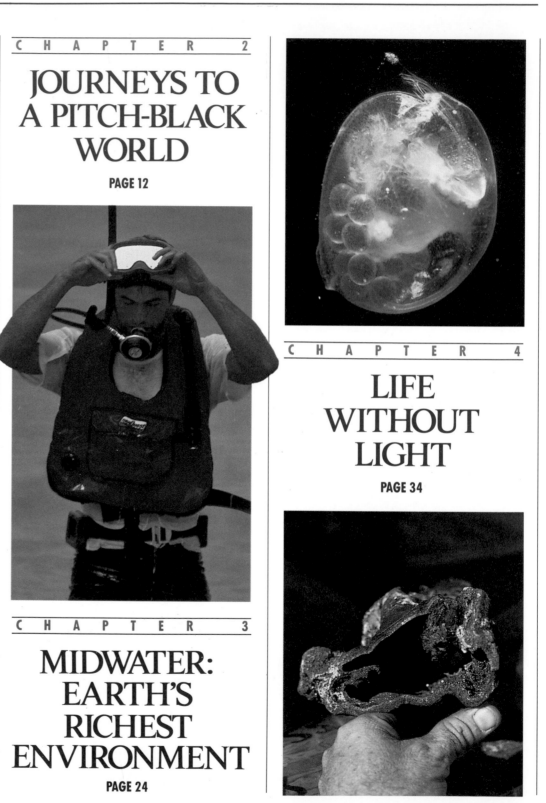

CONTENTS

CHAPTER 1

EARTH'S LAST GREAT MYSTERY

More than a half-century ago, the irrepressible scientist William Beebe—a curator at the New York Zoological Society when he wasn't bushwhacking through unexplored tropical jungles or crawling inside the craters of erupting volcanoes—undertook his bravest and riskiest venture: the first journey to the deep sea.

Taking their lives in their hands, Beebe and a companion stepped inside a small, circular steel chamber equipped with thick, fused-quartz windows and little else. This primitive craft boasted no engine and no means of navigation. The only thing keeping it from sinking like a stone to the distant ocean floor was a cable tethering it to a ship on the surface; through this cable flowed the supply of oxygen that kept the explorers alive. Were this supply interrupted, they would have met with a quick death.

In this 1930 adventure, Beebe was employing the world's first research submarine. He called it a bathysphere and did not know until his first descent whether or not it would burst like a balloon under the immense pressure of the deep sea. But it didn't; the sub and its inhabitants survived the first dive and all subsequent ones, including several to a depth of more than 3,000 feet (915 meters).

So Beebe became the first scientist to view the creatures of the deep sea in their natural habitat. He saw swarms of iridescent jellyfish, odd lantern fish with glowing

In 1930, William Beebe and Otis Barton (previous page) risked their lives by venturing half a mile (0.8 kilometers) below the ocean surface in a primitive submersible, the bathysphere.

Beebe and Barton's explorations were sponsored by research societies, such as the National Geographic Society (right), that were just beginning to explore the unknown world of the deep sea.

tendrils, and many other types he could not recognize. And he grasped something far more fascinating: the intoxicating magic and mystery of the unknown ocean. Writing in *Half Mile Down,* he recalled one of his first dives:

"There came to me at that instant a tremendous wave of emotion, a real appreciation of what was momentarily almost superhuman, cosmic, of the whole situation the long cobweb of cable leading down through the spectrum to our lonely sphere, where, sealed tight, two conscious human beings sat and peered into the abyssal darkness as we dangled in midwater, isolated as a lost planet in outermost space."

Today, we know far more about the ocean than we did a half-century ago. Research vessels crisscross the seven seas, taking water

samples and dredging up the creatures of the deep in nets. Other ships send drills thousands of feet into the earth's crust and dig up solid evidence of the planet's climate a million years ago.

Of course, today's explorers do not have to depend on crawling into an iron bubble in order to visit the deep sea. Submersibles have come a long way since Beebe's day. Now they come equipped with engines that free them of restraining tethers, mechanical hands of extraordinary strength and dexterity, and the most advanced gauges and other recording and measuring devices. Many submersibles today can dive deeper than 3,000 feet (915 meters), to depths of 2.5 miles (4 kilometers) or 13,000 feet (4,000 meters).

Tethered subs still do exist—only they are now robotic vehicles, oper-

ated by powerful computers and designed to follow commands from the surface that enable them to carry out challenging research programs and commercial projects more safely and effectively than ever before.

What we've learned so far is fascinating—and so are the submersibles, satellites, and other devices we're using to unlock the ocean's secrets. But as Bruce Robison, an associate research oceanographer at the University of California, Santa Barbara, explains, we still understand remarkably little about the ocean and its inhabitants. "We know more about distant galaxies than we do about two-thirds of our own planet," he points out. "It'll be decades before we truly begin to understand this, the most complex environment on earth."

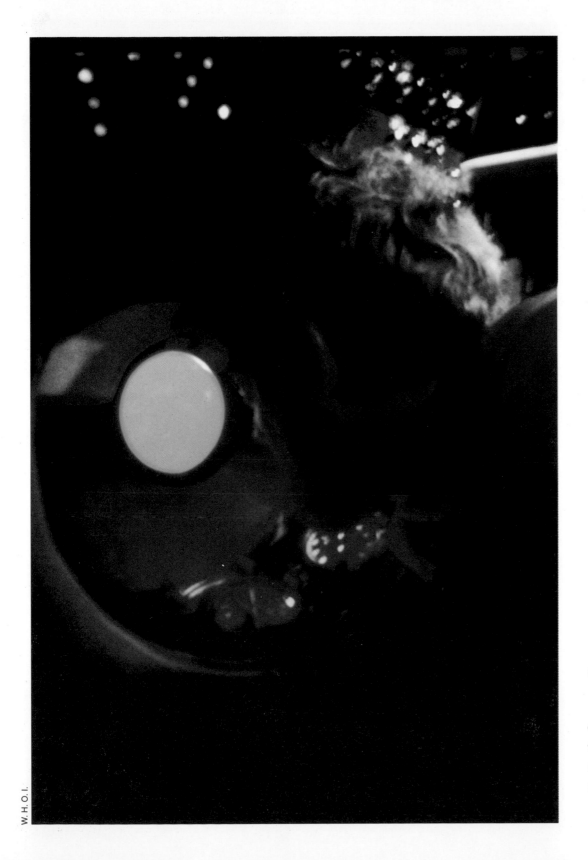

W.H.O.I.

Today's ocean researchers ride in far more advanced subs than did Beebe and Barton (left), but the thrill of exploring the unfamiliar and intricate ocean environment remains the same.

JOURNEYS TO A PITCH-BLACK WORLD

Before William Beebe took his risky voyage in the first bathysphere, scientists attempting to explore the ocean depended on modified one-person diving suits. Some of these involved air hoses attached to large metal helmets; one resembled a floating coffin, within which the diver was lowered into the depths. Not surprisingly, most of these suits weren't very easy to use. Nor did they allow the explorers to descend very far beneath the ocean surface.

Even after Beebe broke through the half-mile (three-quarter kilometer) barrier, years would pass before technological know-how created any craft capable of descending significantly deeper than Beebe's crude model. In fact, the next great leap forward in deep-sea exploration didn't come until the 1960s and 1970s, when *Alvin, Deep Rover,* and other manned submersibles were designed, equipped with powerful engines, complex navigation systems, and—perhaps most importantly—reinforced metal-alloy hulls designed to resist enormous ocean pressure at depths far greater than those reached by Beebe's first explorations.

Among the best known of the new generation of submersibles is the *Johnson-Sea-Link (J-S-L),* owned and operated by the Harbor Branch Oceanographic Institution in Fort Pierce, Florida, a nonprofit organization dedicated to ocean research and exploration. Two of the 22-foot- (7-meter) long *J-S-L*s have

The massive *Johnson-Sea-Links* (previous page)—each submersible the size of a bus—have dropped to half-mile (0.8-kilometer) depths hundreds of times. Pressurized air locks allow divers to leave the sub to do free-floating research and exploration.

The *J-S-Ls'* cameras, strobes, manipulator arms, and other collection and measuring devices (right) obscure the air-tight bubble that holds the pilot and crew.

been in operation since the early 1970s; together, they've made hundreds of dives, both off the coast of Florida and elsewhere.

The two subs have hulls built of welded aluminum, allowing them to descend safely to more than 2,500 feet (760 meters). Unlike Beebe's bathysphere, the 11-ton (9.9 metric-ton) *J-S-Ls* are also equipped with air locks, which allow the divers to leave the subs at depths as great as 600 feet (180 meters), to explore or to retrieve specimens. But it is their auxillary features that make these subs exciting research tools.

For example, both subs have seven-function manipulator arms, each of which can stretch out 8 feet (2.4 meters) and lift up to 150 pounds (68 kilograms). Each arm is equipped with a suction tube; a

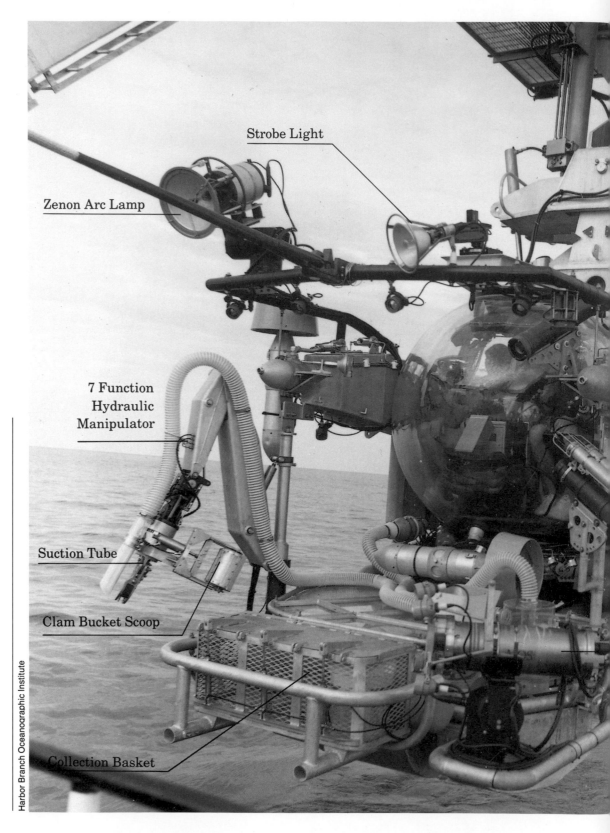

Strobe Light

Zenon Arc Lamp

7 Function
Hydraulic
Manipulator

Suction Tube

Clam Bucket Scoop

Collection Basket

Harbor Branch Oceanographic Institute

W.H.O.I.

Alvin and other submersibles are equipped with surprisingly agile hydraulic arms (left), enabling them to perform complex and delicate deep-sea experiments.

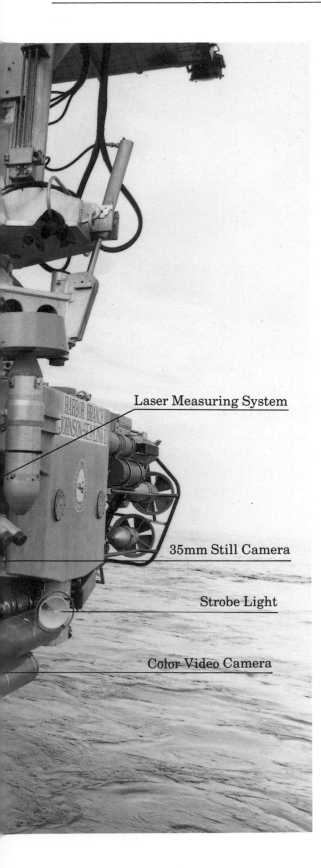

Laser Measuring System

35mm Still Camera

Strobe Light

Color Video Camera

scoop for picking up clams, corals, and other specimens; and an agile hand operated by hydraulic motors. These hands are able to retrieve delicate specimens so carefully that even fragile soft corals aren't broken.

The subs also carry sophisticated photographic equipment. Each sub carries 35mm still cameras (able to take up to 800 frames per dive); 70mm still cameras specially coupled with strobe lights to photograph jellyfish and other nearly transparent sea creatures; and high-resolution video cameras. A work platform at the front of each *J-S-L* holds a wide variety of collection containers, providing storage space for samples gathered by the mechanical arm.

A last unique innovation of *J-S-L*s is the installation of oxygen torches designed to cut through

steel and other hard substances. In 1983, this cutting system was used to sever the anchor chain of the historic Civil War ship the USS *Monitor,* lying off the coast of North Carolina. In this way, the anchor was recovered before it rusted away.

In recent years, *J-S-L*s have participated in dozens of fascinating scientific projects. Since 1984, in conjunction with scientists from the Smithsonian Institution in Washington, D.C., *J-S-L II* has explored the jagged sea floor off the Bahama Islands. During this exploration, it has filmed and captured nearly 120 species of starfish, sea urchins, and related creatures—many of which had never been observed in their natural habitats.

Other projects tackled by the sub include the study of a wide variety of little-known undersea gelatinous creatures off the Bahamas, Bermuda, and the coast of Maine, and research into the private lives of groupers and other fish. In 1984, *J-S-L I* even retraced Beebe's advances, diving into the waters off Bermuda and studying the creatures Beebe had first observed more than fifty years earlier.

While *J-S-L*s are obviously extremely important tools for today's ocean scientist, other submersibles are equipped to undertake even more exciting and difficult studies than those of which the *J-S-L*s are capable. One of the most exciting is *Deep Rover,* designed to peer into the vast, mysterious space known as the midwater.

Deep Rover, designed by Can-Dive Services, a Canadian commercial diving company, is one of the smallest manned subs ever built. Although, like the *J-S-L,* it is certified to dive down about 1,000 feet (304 meters), *Deep Rover* is a mere 10 feet (3 meters) long and 7 feet (2 meters) high, just large enough to hold a single occupant. Its size and quiet engines, says oceanographer Bruce Robison, are its greatest advantages. "Most submersibles are so large and noisy that they make it virtually impossible to watch undersea creatures behaving naturally," he points out. *Deep Rover,*

however, moves very quietly and does not create powerful currents or surges of water that might frighten away small fish or other creatures.

The pilot's cabin of *Deep Rover* is built of clear Plexiglas, allowing 360 degrees of vision for the sub's operator. Sonar devices, depth gauges, and radios enable the pilot to both continuously monitor the environment and to keep close contact with surface personnel. And a pair of manipulator arms holds a video camera and a tactile-feedback mechanism, which allows the pilot to record even the slightest differences in the textures of objects being studied.

The most significant of *Deep Rover*'s many advantageous features, however, is its ability to hang effortlessly in midwater. "Most other submersibles are designed to plunge to the bottom, then perform research there," says Robison. "This sub, however, sets us free to study the largest, and least-known environment on earth: that area that exists between the surface and the sea floor."

A variety of other commercial diving vehicles have been designed by Can-Dive Services, including perhaps the most unusual device of all: the Newtsuit. While primitive

Can-Dive Services Ltd.

Deep Rover (right) is ideally designed for midwater research. The futuristic *Newtsuit* (above) gives divers unprecedented freedom of movement underwater.

individual diving suits predate Beebe and his bathysphere, modern technology has made this extraordinary diving suit a much more important commercial and research diving tool than most previous designs.

The Newtsuit can withstand ocean pressure as deep as 750 feet (230 meters), but what makes it particularly valuable is its revolutionary new joint design. Previous diving suits used oil-filled ball-and-socket joints, which were cumbersome and unresponsive, allowing only five to ten degrees of motion in the arms. The Newtsuit, however, uses stainless-steel rotary joints that permit nearly complete arm and leg motion—allowing far more delicate and precise work.

Despite such newer inventions as *Deep Rover* and the Newtsuit, the best-known submersible of all re-

mains *Alvin,* built by the U.S. Navy in 1964 and operated by the Woods Hole Oceanographic Institution (WHOI) in Woods Hole, Massachusetts. Constructed of titanium—one of the lightest and strongest metals in existence—the sub has the remarkable ability to carry a team of scientists a full 2.5 miles (4 kilometers) below the surface—five times as deep as Beebe ventured and far deeper than either *J-S-Ls* or *Deep Rover* can go.

Like other submersibles, *Alvin* boasts a pair of maneuverable arms tipped with dexterous hands, designed for capturing specimens of unknown deep-sea creatures. It also carries a set of advanced scientific measuring devices, used to sample the water or the sea floor. And, of course, it comes equipped with an array of cameras, both still and video.

The small town of Woods Hole, Massachusetts (right) is bustling with scientific activity. The Oceanographic Institution there (WHOI), founded in 1930 (inset), sponsors dozens of deep-sea research projects each year.

W. H. O. I. W. H. O. I. (inset)

WOODS HOLE OCEANOGRAPHIC INSTITUTION 1930

Alvin in diagram (right) has room for a pilot and two scientists, as well as for cameras, measuring devices, and the complicated workings of its hydraulic arm.

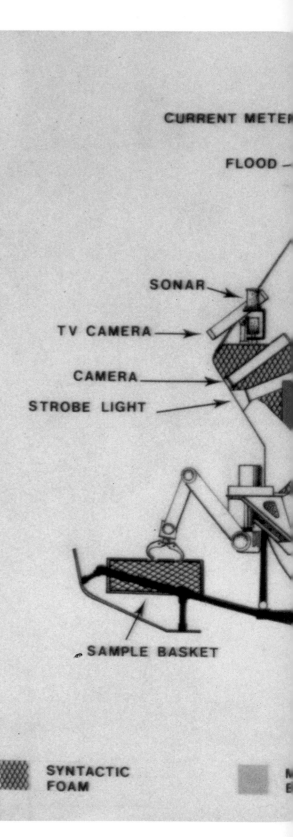

CURRENT METER

FLOOD

SONAR

TV CAMERA

CAMERA

STROBE LIGHT

SAMPLE BASKET

SYNTACTIC FOAM

W.H.O.I.

According to Barrie Walden, WHOI's manager of submersible engineering, *Alvin* became famous soon after it was built, but wasn't widely accepted by scientists until recently. "For a long time, the sub's fame rested on something that happened soon after it became operational," he recalls. "It might never have survived if the Air Force hadn't lost a hydrogen bomb."

The incident referred to occurred in 1966, after a U.S. Air Force B-52 collided with another plane, dropping a nuclear weapon into the Mediterranean Sea off the coast of Spain. For two months, *Alvin* scoured the ocean floor, finally locating the bomb and helping bring it to the surface. "We received a lot of attention for that," says Walden. "And we needed all the help we could get, because back then very few scientists were willing to use the sub for research purposes."

The problem researchers found with using the sub, Walden remembers, was one of dependability. "Why should scientists risk hard-won grant money on a submersible that might sink, or might go out of business in the middle of their project?" These fears were only reinforced in 1968, when the sub fell from its support cables and sank to the bottom, 5,000 feet (1,500 meters) beneath the surface. There it lay for more than a year before being recovered and raised by another submersible. (Surprisingly, it was still in relatively good shape, as were some sandwiches that had been on board when the sub sank. The frigid temperatures and great pressure had kept the sandwiches in nearly perfect condition; the

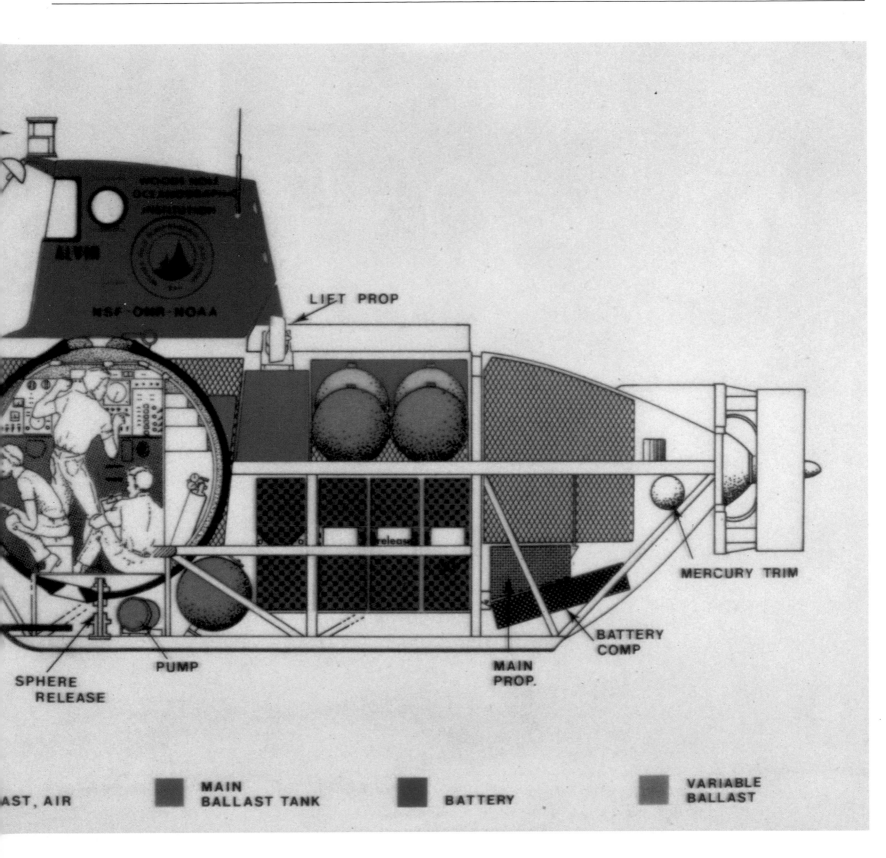

LIFT PROP

MERCURY TRIM

BATTERY
COMP

SPHERE
RELEASE

PUMP

MAIN
PROP.

AST, AIR

MAIN
BALLAST TANK

BATTERY

VARIABLE
BALLAST

bread was a little soggy, researchers reported, but the food tasted fine.)

Despite this lack of early enthusiasm, *Alvin* survived, and today it is in nearly constant demand by researchers for projects throughout the world. Since 1964, the sub has logged more than 1,700 dives. In the past year, traveling aboard the *Atlantis II,* its powerful and fast-moving mother ship, the sub has dove along the coast of the eastern United States, Panama, San Diego, Hawaii, Guam, Yokohama, Japan, and Seattle, Washington.

Such an itinerary is not an unu-sual work load for the vessel, according to Walden. "Most years, *Alvin* stays out at sea for as long as ten months," he says. "Geologists use it to study the composition of the ocean floor; microbiologists capture new types of deep-sea bacteria; chemists look for insights into the chemistry of the earth."

The *J-S-L*s, *Deep Rover, Alvin,* and other submersibles are finally forcing the ocean to reveal some of its mysteries. "It's about time," Robison says. "But we can't get cocky—after all, we haven't discovered even a tiny fraction of what there is to see down there."

The fast-moving *Atlantis II* (left) shuttles *Alvin* and a team of scientists around the world's oceans, often for months at a time.

Alvin (below) was unveiled in 1964, but it has been entirely rebuilt since then. The most important improvement was a titanium hull that allows the submersible to descend 12,000 feet (3,657 meters) beneath the ocean's surface.

MIDWATER: EARTH'S RICHEST ENVIRONMENT

"The only other place comparable to these marvelous nether regions, must surely be naked space it-self . . . where the blackness of space, the shining planets, comets, suns, and stars must really be closely akin to the world of life as it appears to the eyes of an awed human being, in the open ocean, one-half mile [¾ kilometers] down."

This was Beebe's reaction to his record-breaking descent into the frightening and thrilling depths of the midwater in the early 1930s. Here, off the coast of Bermuda, he entered unexplored terrain and caught the first glimpse of a universe of luminescent fish, ghostly jellyfish, and dozens of other creatures. This was a world thriving without sunlight and seemingly without barriers, an amorphous, ever-changing environment unlike any other on earth.

Beebe's heartfelt excitement upon entering this world remains as vivid in his words today as when they were published in 1934. Amazingly, however, we know little more about the midwater—that area that stretches from below the surface of the ocean to the ocean floor—than we learned from Beebe's handful of dives half a century ago. And, according to oceanographer Robison, much of what we think we know is wrong.

"For years, researchers studying the midwater have depended on a single type of research," Robison explains. "You take your boat out, drag a trawling net behind you for

© Bruce H. Robison

a while, then pull it up and see what you caught." In these nets, researchers find an unsurprising variety of fish, shrimp, and other creatures widely considered the most abundant forms of life in the midwater of the ocean.

At the bottom of every net, however, scientists also find a flaccid mass of jelly—the remains of countless jellyfish and of other little-known and insubstantial midwater creatures known collectively as "jellies." By the time they reach the surface, says Robison, these delicate organisms are often damaged beyond recognition. "Even if you could identify some of them, how could you possibly think you were getting an accurate look at the interaction of life in the midwater?" he asks.

Robison himself conducted this type of research in Monterey Bay, an extremely organism-rich environment off the California coast, and found his studies unrewarding and deeply frustrating. "What makes our lack of knowledge even more surprising," he says, "is that the midwater remains by far the largest virtually unexplored area left on Earth."

The awe-inspiring dimensions of this world within the ocean are readily apparent. After all, the ocean covers more than two-thirds of the earth's surface, ranging in depth from shallow coastlines and coral lagoons just a few feet deep to uncharted abysses that may plunge 30,000 feet (9,144 meters) below the surface. Subs certified for only a fraction of that depth find most of the ocean off limits because much of the ocean floor lies 10,000 feet

(3,050 meters) or more below the surface.

Though remote, this vast expanse is not just a "blue desert," says Robison. "To put it simply, most of the earth's animals live in the midwater," he reveals. "Without a doubt, that includes a vast number that have never been seen or identified."

To remedy the glaring deficiencies in our scientific knowledge of this environment, in 1985 Robison recruited five scientists from the University of California, WHOI, and elsewhere. Their goal: to be the first researchers to undertake a comprehensive survey of one midwater area. Their destination was to be Monterey Bay, both because of its richness and because Robison wanted to be able to compare any new results obtained to

© Bruce H. Robison

The mysterious world of the midwater is filled with unfamiliar and intriguing animals. Among the eeriest is the fragile medusa (page 25).

Bruce Robison's dives in the *Deep Rover* captured many of the midwater's most fascinating animals, including this ostracod (above left) and snipe eel (above).

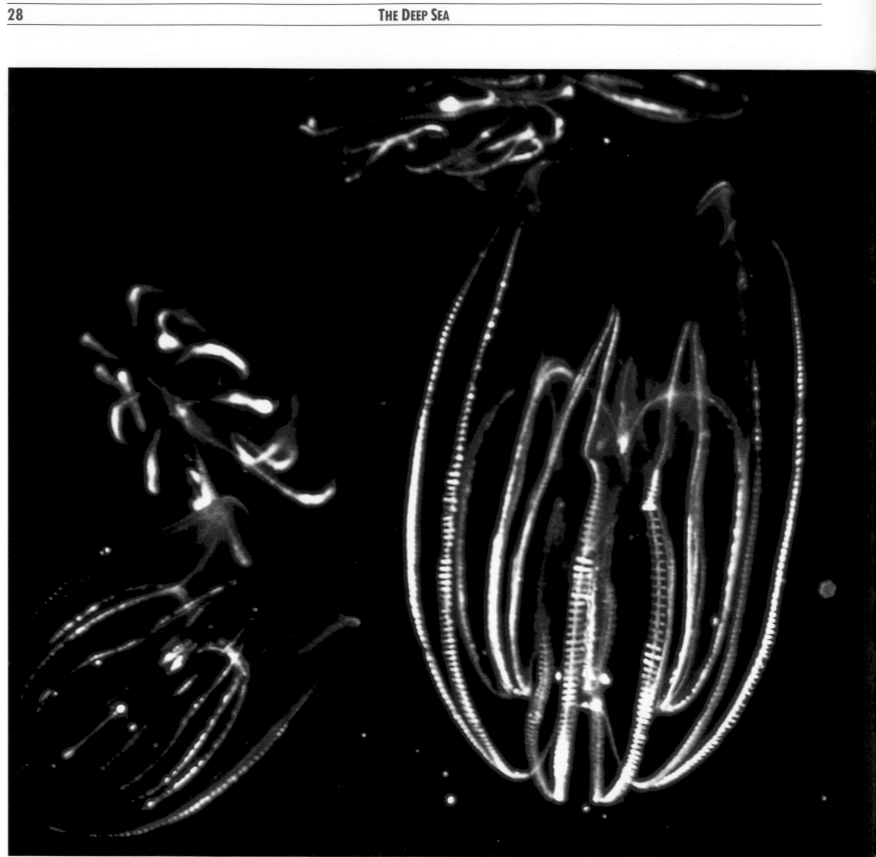

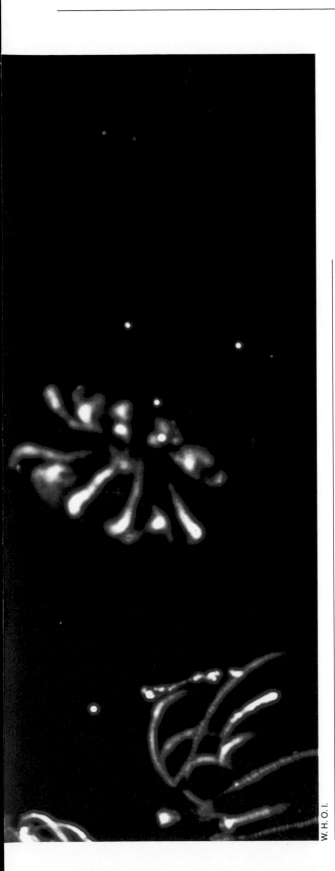

W.H.O.I.

The dominant living creatures in the midwater are the huge masses of "jellies," fragile entities rarely observed before. When disturbed, many glow with cold light.

those he had previously gained using nets in this area.

For this month-long project, funded by a grant from the National Science Foundation, Robison chose *Deep Rover,* designed specifically for work in the midwater. "We knew that simply by going down there, we'd be adding a bias to our studies," he explains. "To get accurate results, we had to minimize the disruptive effect of a large, noisy submersible in that silent world." *Deep Rover,* the size of a small car and with room for just one person, was the perfect solution. (By comparison, the cumbersome *J-S-L*s are the size of a Greyhound bus.)

Armed with video cameras, sampling devices, mechanical arms, and red lights (which don't disturb most animals as much as standard white spotlights), Robison and his team embarked on a series of dives to depths of 3,000 feet (915 meters). The sensation of hanging silently in the midwater, watching a previously unstudied universe of animals drift or swim by, was profoundly moving to Robison and the other scientists. "Being down there alone, seeing that environment for the first time, was awe-inspiring

and a little frightening," recalls Robison. "It was so exciting that all five scientists wanted to go down on every dive. You never knew when you might spot something no one had ever seen before."

Making those dives to the midwater even more spectacular was the presence of a multitude of bioluminescent creatures—animals capable of producing their own light through a chemical reaction inside their transparent bodies. Fireflies are among the few land creatures able to perform this feat; in the darkness of the deep sea, it is a common phenomenon.

The scientific team's bioluminescence specialist was biologist Edie Widder, also of the University of California, Santa Barbara. While aboard *Deep Rover,* she watched bioluminescent fish, jellyfish, and other animals and found, surprisingly, that they usually remained dark until touched by a net attached to the sub. Then they would light up like a display of fireworks.

Attempting to understand why these luminescent midwater creatures light up so rarely, Widder offers varying explanations. "Some may use sudden bioluminescence as a means of startling predators, giving the luminescent prey time to flee," she explains. "Predators may also be hesitant about attacking an animal that may also illuminate *them*, leaving them vulnerable to attack by still larger hunters." Other bioluminescent species, she adds, use light as a sexual display, but only at certain times during the year.

Scientists, however, remain unsure about the possible advantages of bioluminescence to many creatures. "But we do know that it is an extremely important adaptation—or else we wouldn't see it so often," says Widder. "Fish, crustaceans, jellyfish—all share the same ability. With enough time, we'll find out why."

Among the most important discoveries made in Monterey Bay by Robison's team was the realization that jellies—including such little-known creatures as *salps, medusae* and *siphonophores*—are even more abundant than anyone had anticipated. "There are so many of them, and they play such an important role down there, that these jellies are forcing us to reevaluate our picture of life in the midwater," says Robison. "For example, many of them are important predators, while others serve as vast food sources for predatory fish. How could we possibly have realized that by looking at a pile of jelly in a trawling net?"

Richard Harbison, a senior scientist at WHOI, has collected midwater jellies in the Caribbean and off Cape Cod as well as in Monterey Bay. "How many unknown species are there?" he ponders. "We haven't any idea. But every time I go on a collecting expedition, I spend months naming all the previously undescribed animals I bring back."

The discovery of such huge masses of gelatinous creatures has also helped disprove a widely accepted belief about the midwater. "Everyone considers the midwater as having no barriers, no boundaries—and therefore no areas for organisms to hide from each other or establish territories," Robison explains. "But barricades do exist."

Scientists have found, for example, that currents frequently carry plumes of organic matter composed of silt or the crushed shells of dead animals through the midwater environment; these provide at least temporary shelters. Even more importantly, many of the area's unfamiliar gelatinous animals drift through in such vast numbers that they too add a certain structure to the community by creating impermanent walls and hiding places.

Many of the smaller, more vulnerable midwater creatures have two excellent means of cover. The first is to join with others of their species until they form a seemingly huge, invulnerable animal. Siphonophores, a type of midwater jelly distantly related to jellyfish, are quite small. Vulnerable individually, they string themselves together in huge chains that can be more than 30 feet (9 meters) long and thereby discourage predators.

This cooperative life of the siphonophores is even odder than it might first seem. Over generations, some of the members of a chain have evolved to function only as the chain's stinging tentacles, others

W.H.O.I.

WHOI scientist, Richard Harbison, (above) has collected jellies in the Atlantic and Pacific Oceans and in the Caribbean Sea, such as this *Siphonophore calycophoran* (right). He hopes someday to study the unknown midwater creatures that live off the coasts of Africa and Australia.

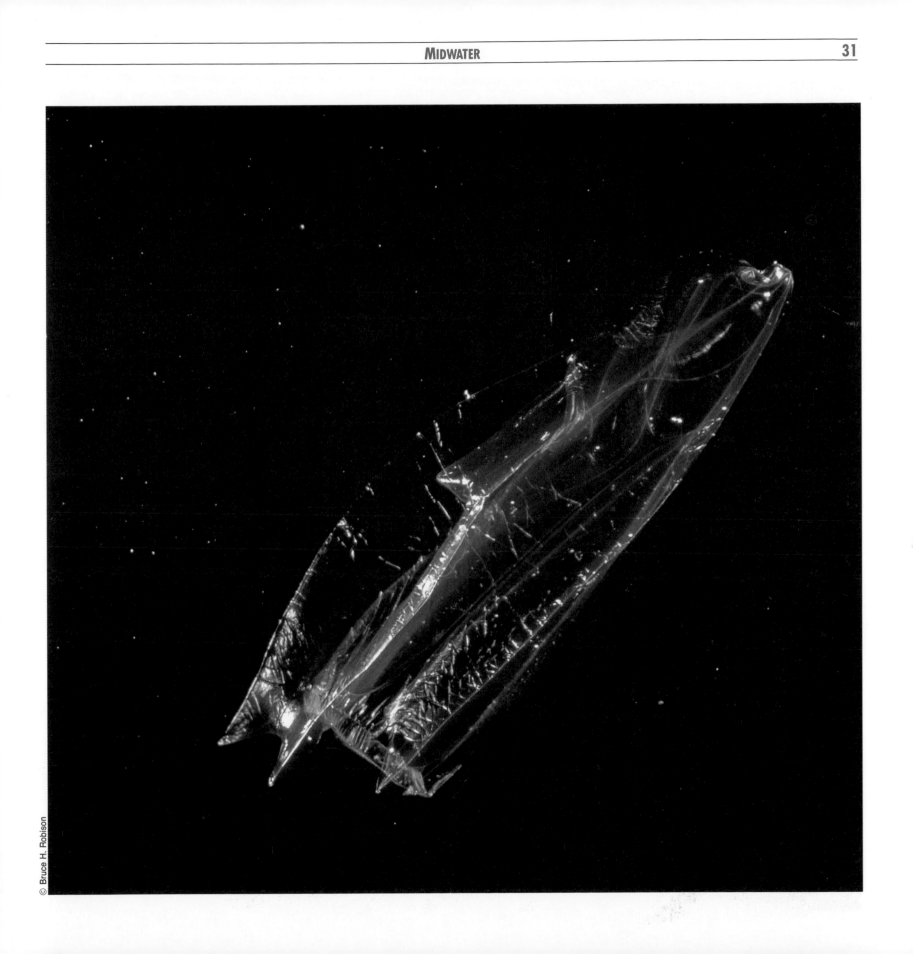

Lou Lehmann, courtesy Can-Dive Services Ltd.

Bruce Robison and the other midwater researchers are eager to use the small, quiet *Deep Rover* (left) for future research.

as its stomach, while still others navigate the group. Only by working together do they serve as a complete animal.

The second survival tactic is to hide behind larger animals that pose no danger; this method, says Robison, also frequently involves the siphonophores. "The chains are so huge, they serve as perfect hiding places for certain small fish and shrimp," he explains. "These animals nestle in the siphonophores' tentacles. They've become immune to the tentacles' poison, so they're safe from larger predators. Plus, they can steal scraps of the siphonophores' prey."

While continuing to review his discoveries from 1985's *Deep Rover* mission, WHOI's Harbison also has plans for several new collecting trips. "I've gone exploring in *Deep Rover, Johnson-Sea-Link, Alvin,* and in scuba-diving gear," he says. "Anything that gets you to the midwater is great. I just wish I could buy a submersible of my own; I'd take it to Australia, Africa—all over the world."

Robison's more focused objective for the future includes two return

trips in *Deep Rover* to the Monterey Bay area. "Since we're sure that midwater communities differ greatly depending on their location, the amount of research left to be done worldwide is nearly incomprehensible," he says. "But there's still so much for us to see in our one little area that I can't wait to get back."

Robison is eager to return to Monterey Bay for another, more philosophical, reason. "The midwater is the only truly three-dimensional environment on earth," he says. "All other creatures are bound by gravity, by permanent barriers, by the earth itself—but not the animals of the midwater. Having never been trained to study a world like this one, we find it very difficult to grasp how profoundly important that third dimension is."

The solution? "The more time we spend down there, a part of that great ecosystem, the easier it becomes to understand," Robison says. "Only by returning and forcing ourselves to toss out conventional wisdom can we find the keys to this greatest of all environments."

LIFE WITHOUT LIGHT

Imagine discovering a fertile green oasis amid the dusty craters of the moon, a bubbling hot spring atop Mount Everest, or a coral lagoon in the Sahara Desert's parched, seemingly endless wastes. In doing so, you will perhaps begin to understand the astonishment of a group of scientists exploring the Pacific Ocean floor in *Alvin,* 8,000 feet (2,500 meters) below the surface in 1977.

Their visit was the eventual result of some odd data recorded earlier by Woods Hole scientists mapping the temperatures of the deep sea from the surface. In an area near the Galápagos Islands in the Pacific, they found higher water temperatures than they had expected. This meant that some sort

of hydrothermal activity was going on, releasing heat from the earth's core into the water.

Such unexpected temperature variations helped the researchers pinpoint the location of a spreading center—a place where the earth's surface is separating and new crust is being formed. Such spreading centers are found along the Mid-Ocean Ridge, an immense underwater mountain range that stretches for 40,000 miles (64,000 kilometers), or nearly a quarter of the earth's surface. According to WHOI biologist J. Frederick Grassle, "Much of the Mid-Ocean Ridge is volcanically active, which means that new sea floor is being formed there." Depending on the amount of volcanic activity, the spreading

W. H. O. I./Rod Catanach

W.H.O.I.

WHOI geologist, Robert Ballard, (above) and biologist, J. Frederick Grassle, (far right) were among the first men to discover the multitude of mussels and other strange creatures (right) that thrive 8,000 feet (2,500 meters) beneath the ocean's surface.

Alvin (previous page) has carried scientists to the ocean floor on more than 1,500 dives; the most exciting were undoubtedly those that first explored the mysterious deep-sea vents.

W. H. O. I./J. Frederick Grassle

W. H. O. I.

rate may range from 2.5 inches (6 centimeters) a year to as much as 7 inches (18 centimeters).

To gather further information about the spreading center located near the Galápagos, in 1976 scientists from Woods Hole and elsewhere dispatched more research vessels to the site. Though *Alvin* was not yet available for this voyage, scientists used *Deep Tow,* a small sub pulled by a ship and designed to map the ocean floor using high-speed photography.

Deep Tow's passage over the spreading center revealed the next intriguing find there: the presence of large numbers of clams in the area, their shells gleaming in *Deep Tow*'s spotlights. "Even this interesting discovery was not considered important enough to warrant sending a biologist on *Alvin*'s first visit to the area," recalls Grassle. When WHOI geologist Robert Ballard and others rode the tiny submersible to 8,000 feet (2,500 meters) to take a look, however, they came upon an almost unbelievable sight.

Instead of an empty gray expanse, as expected, the *Alvin* team saw gaping fissures in the sea floor, through which spewed vast quantities of hot (nearly 700°F [370°C]), murky water. But it was the life around these deep-sea vents that truly astonished Ballard and made Grassle and other scientists back on the surface eager to visit the site.

Here, in this pitch-black, barren environment, was living a vast community of remarkable—even unrecognizable—creatures. Closest to the vents were large yellow mussels in sprawling clusters. Farther away, the explorers saw a multitude of huge white clams, whose gleaming shells had been captured by *Deep Tow*'s photographs.

Others of the vent inhabitants were far stranger. Suspended above the hot-water fissures were flower-like yellow organisms—obviously some kind of animal but resembling nothing so much as dandelions gone to seed. Nearby, pink fish hung, heads downward, over the vents, while delicate, free-swimming worms twirled past *Alvin*'s windows in a twisting dance.

The most dominant presence in this spreading center population was that of one particularly strange creature. Near the vents, standing upright like a grove of birch trees, were giant reddish pink worms encased in paper-thin white sheaths. A large colony of these worms waved in the current; occasionally one would thrust a bright pink plume into the water, then pull it back inside its sheath. Some of these worms were more than 8 feet (2.4 meters) tall. Ballard—and the biologists who reviewed the photos from the first expedition—had never seen anything quite like them before.

These discoveries were, as Grassle points out, unexpected and exciting for several reasons. "We know that most animals living on the sea floor are relatively small, depending on food that gradually filters down from the surface," he explains. "So, the moment we heard that large animals had been spotted at the Galápagos Spreading Center, we knew that they had to be relying on some other food source."

At about the same time, Grassle and his colleagues also received reports that the water around the vents was very rich in hydrogen sulfide, or sulfur, a chemical that is poisonous to most living creatures. Certain types of bacteria, however, are known to thrive on sulfur, using it to create food. "So it wasn't very difficult to make the connection," Grassle recalls. "We guessed almost immediately that these bacteria might be the key to the vent community."

Countless other details of life at the vents, however, could not be understood without further exploration. In 1979, therefore, teams of biologists (including Grassle himself), microbiologists, biochemists, and other scientists began to visit the Galápagos vents aboard *Alvin*. They soon discovered that their predictions concerning the presence of bacteria were correct: Sulfur-loving microorganisms could be found in great abundance at the vent openings, in concentrations hundreds of times higher than elsewhere in the deep sea.

The discovery of bacteria at hot-water vents 1.5 miles (2.4 kilometers) down may not, at first, seem cause for great excitement. But for Grassle, and every other scientist involved, the proof of their theory was thrilling, because it turned upside down many previously accepted "facts" about the world we live in.

It had long been a basic assumption of biology that no living creature can survive without light. The reason: All living things depend on plants for food (even meat-eating creatures eat plant eaters), and

W. H. O. I./Dudley Foster

Perhaps the strangest of the ocean creatures recently discovered are Riftia, the giant tube worms (above and left). Measuring up to 8 feet (2.5 meters) in length and reddish-pink in color, the worms are only found near deep-sea vents.

W. H. O. I./Kathleen Crane

Agile and quick, *Alvin* was able to capture the giant white clams, mussels, and many other vent creatures (left). Most of them have never been studied before.

Here is a close-up view of the masses of bacteria that nourish the mussels, clams, tube worms, and other vent animals (right).

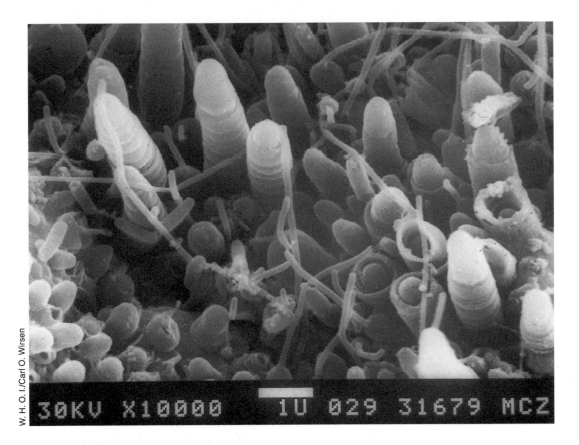

W. H. O. I./Carl O. Wirsen

30KV X10000   1U 029 31679 MCZ

plants require sunlight for photosynthesis, the process by which they manufacture their own food. No living thing, scientists believed, could survive in a world without plants. Even most of the creatures that populate the ocean floor, though living in a world of total darkness, are still dependent on food filtering down from the water's surface; food that grew in the light of the sun.

Yet here, at 8,000 feet (2,500 meters), Grassle was exploring a world he would never have guessed existed, a world not only blanketed in perfect darkness, but one where sunlight *simply didn't matter*. Photosynthesis wasn't necessary for these giant worms, mussels, and clams to survive: The masses

of bacteria discovered in these dark waters, busily converting hydrogen sulfide and oxygen into organic nutrients, were quite able to survive without the sun. And, by doing so, they were fulfilling the same role— the first link in the food chain—as that fulfilled by photosynthesizing plants everywhere else on earth.

It is possible, therefore, that if the sun were suddenly to be extinguished, all life on earth would soon perish—except for these creatures in the dark waters surrounding the warm vents of the deep ocean.

Having found so much life thriving at the Galápagos Spreading Center, Grassle and others were determined to learn as much as they could about the members of the

vent community. Capturing specimens with *Alvin*'s grasping hands and with collecting jars, the scientists soon discovered that some of the vent animals have relatives elsewhere in the deep sea. "For example, the vent mussels and clams are both related to species found far from any vents," says Grassle.

Others of the vent creatures, however, were proven to be just as unique and alien as first thought. Perhaps the most interesting of all were those remarkable 8-foot (2.4-meter) tube worms, which were christened *Riftia* and are thought to belong to their own family. They are larger and more spectacular than even their distant relatives, and thus far have been seen only around the Galápagos vents.

These yellowish creatures (right) were nicknamed "dandelions," but they are actually related to jellyfish. Crabs (far right) are also abundant near the Galapagos vents and easy to attract.

W. H. O. I./Robert Hessler

W. H. O. I./Robert Hessler

Among other intriguing animals observed were two species of crabs spotted near the vents. These crabs, now known to belong to a new superfamily (which means they are not closely related to anything ever seen before), are thought to be scavengers, as they readily come to bait dangled from *Alvin*—making them easy to catch compared to other vent creatures.

The beautiful "dandelions" seen floating above the vents are now thought to be kin to the Portuguese man-of-war jellyfish. They are able to swim freely, but spend much of their time attached to rocks by their tentacles, which are thought to trap food particles from the water, much as do these of more familiar jellyfish. Oddly, at one Galápagos vent where scientists found no hydrogen sulfide in the water, these animals are the dominant species. No one is sure why.

Despite much effort, scientists aboard *Alvin* have been unable to capture the pink fish frequently seen near the vents, but they believe the species to be part of a known family of deep-sea fish. "We've never seen it feeding, and have been unable to attract it with bait, so its behavior and habits are still rather mysterious," says Grassle. "It's a very interesting fish."

While this creature has evaded capture, most of the others found at the Galápagos vents have been brought to the surface for study. Of the many questions scientists have hoped to answer by analyzing the vent animals in the laboratory, perhaps the most important is also the most fundamental: How and, in some cases, what do they eat?

In many cases, the answer seems fairly clear. Limpets, anemones, and other organisms may simply filter the nutritious bacteria directly from the water near the vents, while the crabs seem certain to depend on scavenging for their meals. A gray, snakelike fish seen at the vents is thought to be a predator, finding its prey at these rich hunting grounds.

Many other creatures, however, live further from the rich source of bacteria at the vents, and their feeding habits are not so easily surmised. The large white clams, for example, are never found directly next to the vents—and, of course, they are unable to move very far in search of food. How, then, do they survive?

The answer, researchers found, is a clever example of the cooperative strategy known as symbiosis. Vast quantities of bacteria live on the clams' gills—far more than could be expected from the relatively sparse numbers found in the surrounding water. Scientists therefore believe that the bacteria use the clams as a home, complete with a supply of nutritious water passing through. The clams, in turn, eat some of the bacteria, but not enough to threaten the population.

A similar, but even more interesting, feeding technique is practiced by *Riftia*, the giant tube worms that have come to embody all the strangeness of this pitch-black underwater world. At first, Grassle remembers, scientists were confounded by the worms' remarkable physiology. "We had no idea how they could feed," he says, "because they have neither a mouth nor a gut." Unlocking this mystery, scientists dissecting the creatures

W. H. O. I./Kathleen Crane

Scientists have not been able to capture this strange, pink fish (above), often seen around the Galapagos vents. So far, they have also never seen it feeding, but only hanging, head-down, over the vents mysteriously.

Crabs and tube worms (left) thrive in the pitch-black, seemingly barren world of the deep-sea vents of the Galapagos.

found that where the intestines should be, were pockets of bacteria instead.

Once again, symbiosis was the key. Apparently, the worms ingest sulfur-rich water through their pink plumes. This diet of hydrogen sulfide supports the bacteria colonies inside the worms, which survive by eating some of the bacteria, which then provide food for the worms. Exactly how the bacteria colonize inside the worms is just one of the additional questions that has kept researchers busy since their initial discovery.

After several visits to the Galápagos vents, Grassle and other scientists felt they had at least begun to understand something of the dynamics of life in this isolated world. (They had also become comfortably familiar with the area, giving different regions such names as "Clambake" and "Garden of Eden.")

With study of the Galápagos vents so well underway, scientists began to wonder if other vent communities also existed. After all, spreading centers were known to be common throughout the volcanically active Pacific—so it would make sense that animals would also take advantage of those other warm, sulfurous waters.

Confirming scientists' expectations, subsequent explorations in an area known as the East Pacific Rise off the coast of Mexico turned up several more vents with their own communities, as did dives to the Guaymas Basin in the Gulf of California. In recent years, vents have also been found far farther north in the Pacific—including

W. H. O. I./Rod Catanach

Some vents are shaped like odd, stone formations (above). These "black smokers" (right), seen at some Pacific vents, spew mineral-rich hot water into the cold ocean.

some where hot water spews out of stone formations dubbed "black smokers." Some of the animals seen at each site are the same as those found at the Galápagos vents, but others, including smaller relatives of the giant tube worm, have been new discoveries.

The most exciting of all these new vent discoveries, according to Grassle, have also been among the most recent. Unusual vent communities have been found in the Atlantic Ocean, as have others off the coast of Florida that resemble hot-water vent communities but seem to thrive in an area where there are no hot-water vents at all. Instead, at the sites of these communities, cold, nutrient-rich water seeps up from the sea floor like

drinking water from an artesian well.

"We weren't surprised to find vents in the Mid-Atlantic Ridge, but we didn't expect to see such different animals there," says Grassle. "For example, there are vast swarms of shrimp that have never been seen in the Atlantic before. They simply cover the rocks at these vents, in uncountable numbers."

The cold-water seeps in the Gulf of Mexico off the coast of Florida are even more exciting—and puzzling. Scientists had always assumed that vent communities depend on hydrothermal activity, the pumping of hot water into the ocean. But here they found mussels, clams, and tube worms living

Newly discovered Atlantic Ocean vents support uncountable swarms of shrimp, a type never before seen in the Atlantic.

W. H. O. I./John Edmond

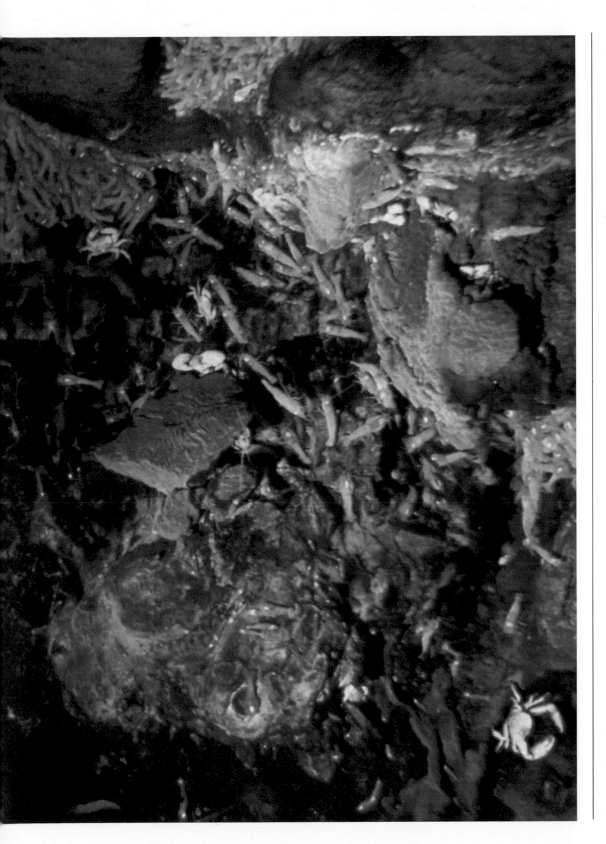

quite comfortably in the absence of hot water. "This discovery shows that the number of environments where we can find these creatures may be far greater than we ever guessed," Grassle points out.

While searches for other vents are certainly being planned, Grassle feels that researchers still have many questions to answer about the vents already found. He wonders, for example, why animals from elsewhere in the ocean are so rarely seen at the vents. "A few things do wander in occasionally," he says, "but most creatures from the rest of the deep sea seem to avoid the vents. For example, rock surfaces elsewhere are usually covered with corals, sponges, brittle stars, and other animals. We think there's an ecological reason why these don't occur at the vents—possibly because of high concentrations of toxic metals in the water."

Even more intriguing puzzles are how vent creatures reproduce and how they manage to colonize new vents that may be many miles of frigid ocean away. "It's important to remember that, because of the unpredictability of undersea thermal activity, the life span of each individual vent may be no more than a few decades," Grassle points out. "And when the hot water stops pumping, the community of animals there cannot survive."

Already, researchers aboard *Alvin* have noted changes in hydrothermal activity at several vents, and scientists have also spotted vast beds of clam shells at various sites—evidence of dead vents. "Clams' shells dissolve within fifteen years, so we know that these

Dead-clam shells are a sobering reminder that the life of a vent—and of the animals it supports—can be quickly extinguished.

vents supported life until recently," says Grassle. "The challenge is finding out how the creatures—or their offspring—manage to find and colonize new vents, especially such sedentary creatures as *Riftia* or the mussels."

One species seems to have adapted to the transient nature of the vents in a unique way. The large white clams so abundant near the vents are able to grow to maturity at an extraordinary rate—up to five hundred times as quickly as their more familiar relatives. By maturing so quickly, the clams are able to produce vast numbers of free-swimming larvae during a life that may be cut short by the death of the vent itself. Most likely, these larvae drift until reaching a new active vent, where they settle and begin to grow.

Until recently, scientists thought that the clams' offspring (as well as those of other vent animals) must have to travel vast distances to reach new vents, somehow journeying through the frigid ocean until they found a new home. Now, however, it seems that the distances between vents may not be so large.

"There are many vents in a single active area," says Grassle. "So, although one vent may die out, others will survive."

Far more must be learned before researchers feel they have full knowledge of the life of the vent animals. For example, although both the adult tube worms and their eggs have been studied, the larvae have never been seen—and the same holds true for many other vent animals.

In the future, research efforts will surely continue to grow. *Alvin,* for one, will be returning to the Guaymas Basin vents and certainly to others soon thereafter. Researchers will no doubt continue to seek out and study newly discovered vents also, like those recently reported found in the waters off Japan.

"What makes this research so unusual and so exciting is how many scientists from different fields are fascinated by the vents," Grassle says. "Geologists, biochemists, microbiologists, biologists—we all have so much more to learn about what makes these communities tick."

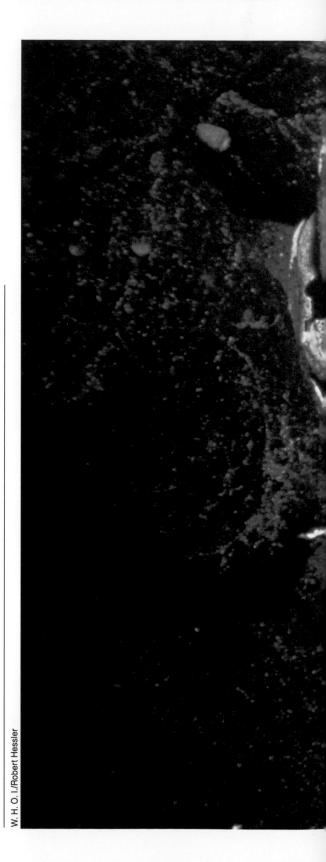

W. H. O. I./Robert Hessler

ARGO, JASON, AND OTHER ROBOT SUBS

Alvin, Deep Rover, and other manned submersibles have made the deep sea far more familiar and comprehensible than ever before. By revealing the beauty of mid-ocean life and delving into the mysteries of the hot-water vent communities, these small subs—and, of course, the scientists who operate them—have opened our eyes to the vivid world that exists beneath the forbidding surface of the ocean.

As Dana Yoerger of WHOI's Deep Submergence Laboratory (DSL) points out, however, such manned vehicles as *Alvin* pose problems when it comes to exploring the deep sea. Their most serious limitations: they are very slow-moving and lack endurance,

being able to stay down only a few hours at a time. "*Alvin* and other manned subs are simply not designed to be surveying vehicles," Yoerger says. "They don't cover enough ground to be effective."

Yoerger stresses that speed is essential if researchers are ever to get a clear view of the enormous expanse of the ocean bottom. "With all the mapping we've done in recent years, if you look at a contour map of the sea floor, you'll still see large areas with no contours. That doesn't mean the area's flat—it means no one's ever looked at it." *Alvin,* whose every dive costs a great deal of effort and money, isn't up to the task.

The risks involved in research with manned submersibles also

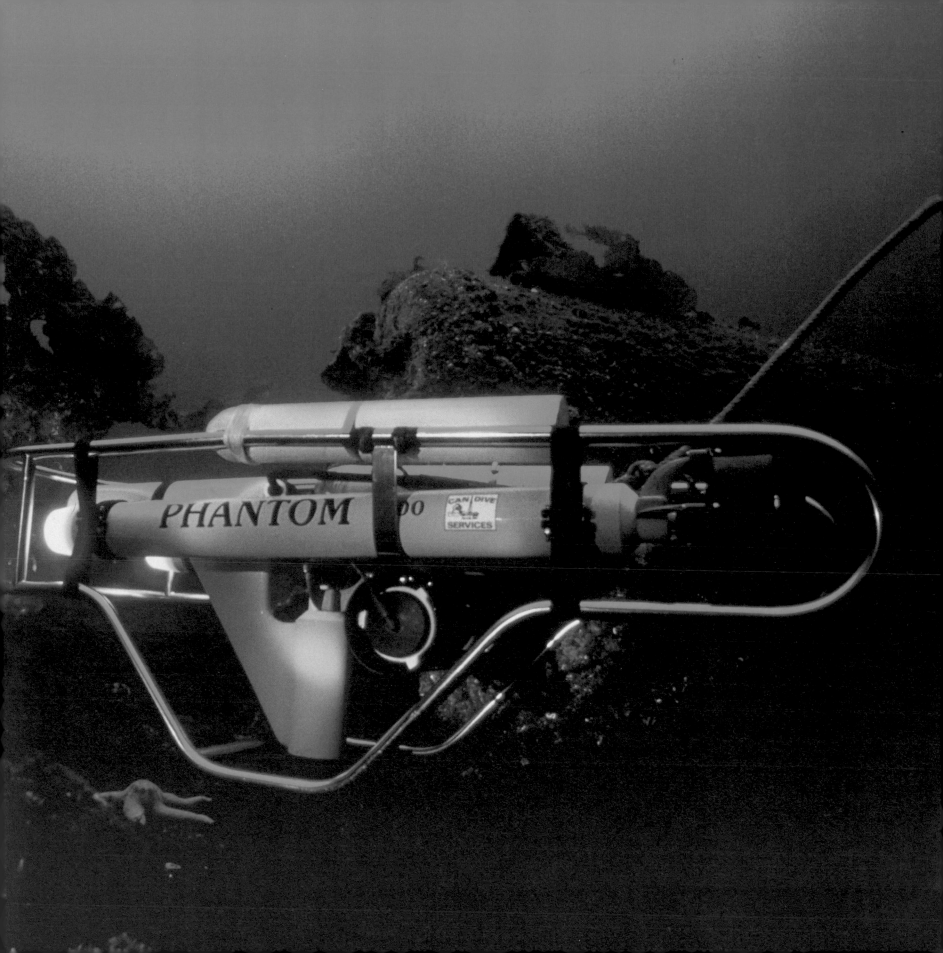

Unmanned subs, like this tiny craft (previous page), are performing an increasingly significant amount of undersea research and industrial work.

W. H. O. I.

ANGUS (right), a sturdy "camera on a rope," can journey 20,000 feet (6,100 meters) below the ocean's surface. It can take up to 3,200 photographs per dive.

limit their capabilities. There are also regions of the ocean waters into which it is simply too dangerous for many submersibles to venture. The ocean floor is no smooth and sandy beach, but is frequently rocky and strewn with rubble. Subs have become entangled and trapped in this difficult terrain, endangering—and sometimes even costing—the lives of the crew.

The demand for safe, efficient, and wide-ranging undersea vehicles not subject to the limitations of manned subs is now strong. The U.S. Navy (a major funder of submersibles) needs such vehicles both for mapping and for the search for and close inspection of sunken ships or downed aircraft. And, of course, scientists at WHOI and elsewhere have an equally vital need for subs that can perform many tasks more safely and less expensively than *Alvin*.

In recent years, unmanned submersibles, generally operated from ships on the ocean's surface, have begun to play an increasingly important role in undersea exploration. While several types exist, the most widely used are really no more than high-powered cameras and sonar equipment towed along above the sea floor.

The first and perhaps the most famous of these search-and-survey systems is *ANGUS* (*Acoustically Navigated Geological Underwater Survey*), built in the late 1970s. *ANGUS* is designed to work primarily in rugged volcanic terrain

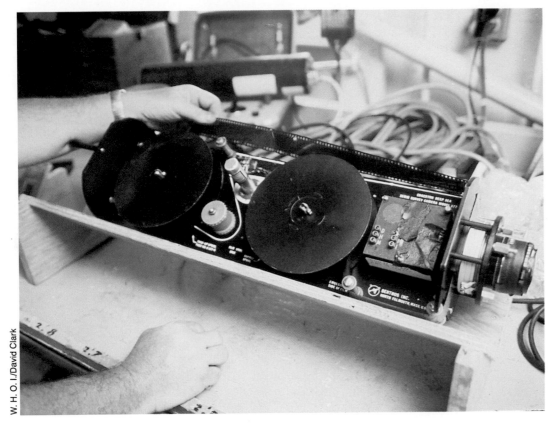

W. H. O. I./David Clark

ANGUS' cameras (left) must withstand enormous pressure, frigid cold, and other harsh conditions.

in depths of up to 20,000 feet (6,100 meters)—far deeper than *Alvin* or any other manned submersible can venture and within reach of more than 90 percent of the ocean floor. It is attached by a cable to a surface ship and has a heavy-duty frame that enables it to survive head-on collisions with treacherous rock outcroppings without damage.

ANGUS is equipped with three 35mm color cameras, each of which is able to take up to 3,200 photographs per dive and which can photograph a section of sea floor 200 feet (60 meters) wide. Working alongside these cameras is a system that monitors variations in water temperature, as well as sonar equipment that keeps the operators on the surface aware of the

vehicle's cruising altitude.

Unlike many survey crafts, *ANGUS'* strobe lights, used to illuminate the bottom, keep the sub in constant visual contact with the sea floor—giving scientists their first clear view of huge expanses of previously unexplored terrain.

A smaller 1981 variation on this system, *Mini-ANGUS,* can be equipped with one- or two-color cameras and other instruments. *Mini-ANGUS'* greatest recent moment of glory may have been when it was employed by the Israeli government a few years ago to scan an undersea weapons range for unexploded bombs.

While *ANGUS* and *Mini-ANGUS'* designs and capabilities may seem comparatively simple,

the challenges facing their designers were great. "It's easy to say, 'Oh, that sub's just a camera on the end of a rope,'" says Yoerger. "But it's important to remember that the vehicle must function under extremely harsh and unpredictable conditions."

These challenges—and many others—also had to be taken into account in the design of one of the Deep Submergence Lab's most ambitious projects: the creation and building of *Argo,* a far more complex and advanced survey vehicle. This towed submersible was responsible for finally locating the *Titanic*—in 1985, in its first open-sea test—after others had spent decades searching for the great wreck. Yoerger and others believe that *Argo* will enable researchers to put together a clearer picture of the deep-sea floor than would be possible with any other survey vehicle.

Developed with funds from the U.S. Office of Naval Research, *Argo*, like *ANGUS*, is tethered to a research ship on the surface—in the case of the search for the *Titanic*, the WHOI vessel *Knorr*. Also like *ANGUS, Argo* has been designed to withstand rough terrain, particularly the rugged surface of the Mid-Ocean Ridge, the world's most massive and least-known mountain range. But *Argo* is also equipped with far more delicate and advanced systems than that other "camera on a rope."

For example, *Argo*'s "rope" is actually a steel-armored cable, only two-thirds of an inch (1.6 centimeters) in diameter yet capable of supporting 35,000 pounds (15,855 kilograms). It is also strong enough to withstand the great pressure of the ocean 20,000 feet (6,100 meters) down while still transmitting complex information back to the surface.

Argo is equipped with powerful strobe lights and film and video cameras that gaze ahead, to the side, and below the submersible. The color video cameras send pictures instantaneously to researchers aboard the *Knorr*, enabling *Argo*'s operators to avoid rock outcroppings and other obstacles while also allowing scientists to scan the ocean floor at the actual moment the sub is passing by, rather than waiting to process film. Also built into *Argo* is an advanced sonar system, which uses sound to draw a

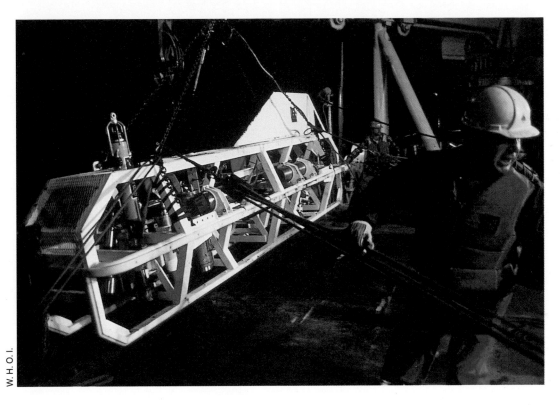

Argo's sturdy metal frame (above), designed to withstand collisions with undersea mountains, protects the sub's complex network of cameras and other instruments.

Argo, an unmanned sub attached by a cable to the surface (left), can carry cameras and other instruments 20,000 feet (6,100 meters) below the surface.

clear picture of the previously unseen mountain ranges, rocky valleys, and other fascinating contours of the deep sea.

The video system is a major advance, according to Yoerger, but is not without drawbacks. "Scientists think it's wonderful that they can sit comfortably aboard the *Knorr,* watching the ocean floor—but video images simply don't have the resolution and dynamic range that film has. Even with *Argo*'s strobe lights, the images we get from the bottom are low contrast, with everything either very light or very dark. Video simply has trouble recording these images clearly—and we haven't found any way of enhancing the images substantially."

Solutions to these problems, however, are already on the horizon. Among the most exciting prospects are filmless, solid-state cameras, which have been under development for years. These would record images with the definition of film, but would transmit the images immediately up the cable, where computers could both enhance them and analyze them.

Another important development, which *Argo*'s designers are looking forward to, is the replacement of the vehicle's tether with fiber-optic cables. Fiber optics, which transmits information via minute pulses of light, can effortlessly handle nearly limitless amounts of data—far more than today's standard copper coaxial cable. The technology is already widely used in communications and computer processing, and is without doubt the technology of the future for submersibles too.

Though they are being developed

This artist's rendering (right) shows the future of deep-sea research: *Argo* will cruise above the sea floor, scanning the bottom with cameras and sidescan sonar, while *Jason* will venture downward for a look at any interesting objects.

W. H. O. I./Stefan Masse

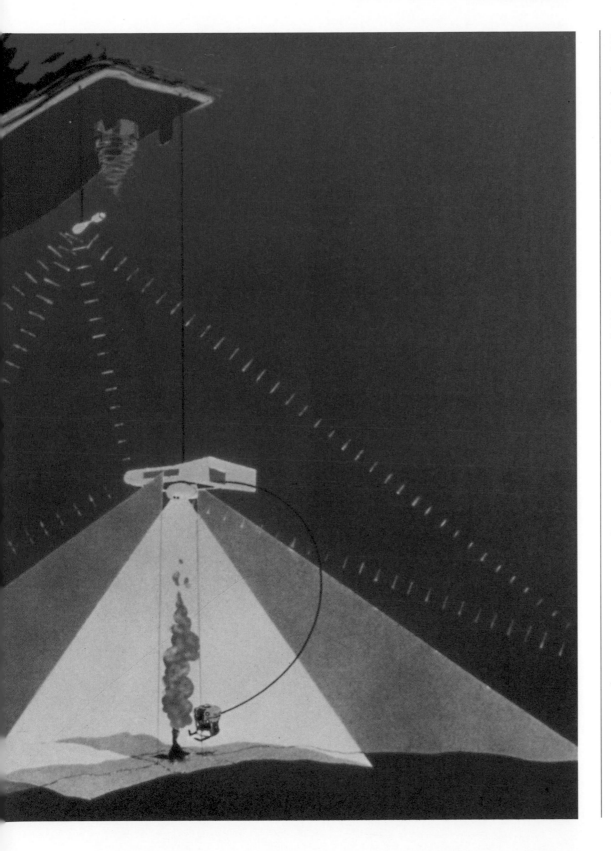

for greater strength, fiber-optic cables are not yet able to withstand the stresses submersibles such as *Argo* require. "We get advertisements from companies demonstrating how strong and durable fiber-optic cables are by showing a tank running over one," says Yoerger. "But that's nothing compared to the stress a cable goes through while supporting a heavy vehicle at great depths. We haven't yet come up with a cost-effective cable that's strong enough—but we will." When they do, *Argo* and other submersibles will be able to supply more information about the Mid-Ocean Ridge and other ocean terrain.

Even without these awaited technological advances, *Argo* has had considerable success. In 1986, following its discovery of the *Titanic,* *Argo* visited the East Pacific Rise off Mexico, the site of several of the most important hot-water vent discoveries. Under the leadership of Ballard, *Argo* succeeded in mapping far more of this volcanically active area than had previously been explored. Its cameras and sonar spotted many new hot springs as well as sites of recent volcanic eruptions.

Despite these early triumphs—and without doubt many more to come—*Argo*'s designers at the Deep Submergence Lab consider their job only partly done. *Argo* was never intended to cruise the ocean floor alone. It was conceived to be accompanied by a small robotic sub named *Jason,* a beetlelike computerized vehicle tethered to *Argo,* much as *Argo* is tethered to the *Knorr.*

Argo's job is to survey and map;

Jason, Jr. (right), directed by scientists aboard *Alvin*, snaked along the *Titanic's* corridors in a remarkable maiden voyage. Soon, the sub will be replaced by *Jason*.

Jason will function as a highly maneuverable robot for close-up inspection and scientific experiments. With *Jason* under tow, scientists scanning the sea floor through *Argo*'s video cameras will be able to send commands down through the cable when they spot something interesting, ordering *Jason* to detach itself and investigate.

The little sub will be equipped with high-quality color "vision" and—most importantly—delicate and versatile manipulators much like *Alvin*'s. Its small size will enable the sub to operate in extremely rugged terrain and in such confined areas as shipwrecks, while its cameras and grasping hands will allow it to capture images and specimens unlike any ever seen before. And *Jason* will perform these tasks quickly, less expensively, at greater depths, and far more safely than manned submersibles ever could.

Jason isn't ready yet, but *Jason, Jr.* (usually called *JJ*), a small prototype operated from *Alvin*, has already established a name for itself. The robot is 20 inches (50 centi-meters) high, 27 inches (68 centimeters) wide, and 28 inches (71 centimeters) long and is equipped with delicate thrusters for maneuvering and with high-quality color video and still cameras. This little box was built to test components for the larger sub. "We see *JJ* as a stepping-stone toward *Jason*," says Yoerger.

Jason, Jr.'s maiden voyage was, to the public, a smashing success. Entering the wreck of the *Titanic*, the little robot ascended staircases, floated through ballrooms, and sent back dozens of spectacular pictures of the great ship. For the people at the Deep Submergence Lab, however, enthusiasm was more tempered. "It worked," says Yoerger, "but we also encountered a variety of problems, particularly with the motors."

Finding the problems is the first step toward solving them, of course, and none of those that appeared during *JJ*'s jaunt seem insurmountable. The robot's future, nonetheless, is undetermined. "We'd love it if scientists in *Alvin* decided they wanted to keep using

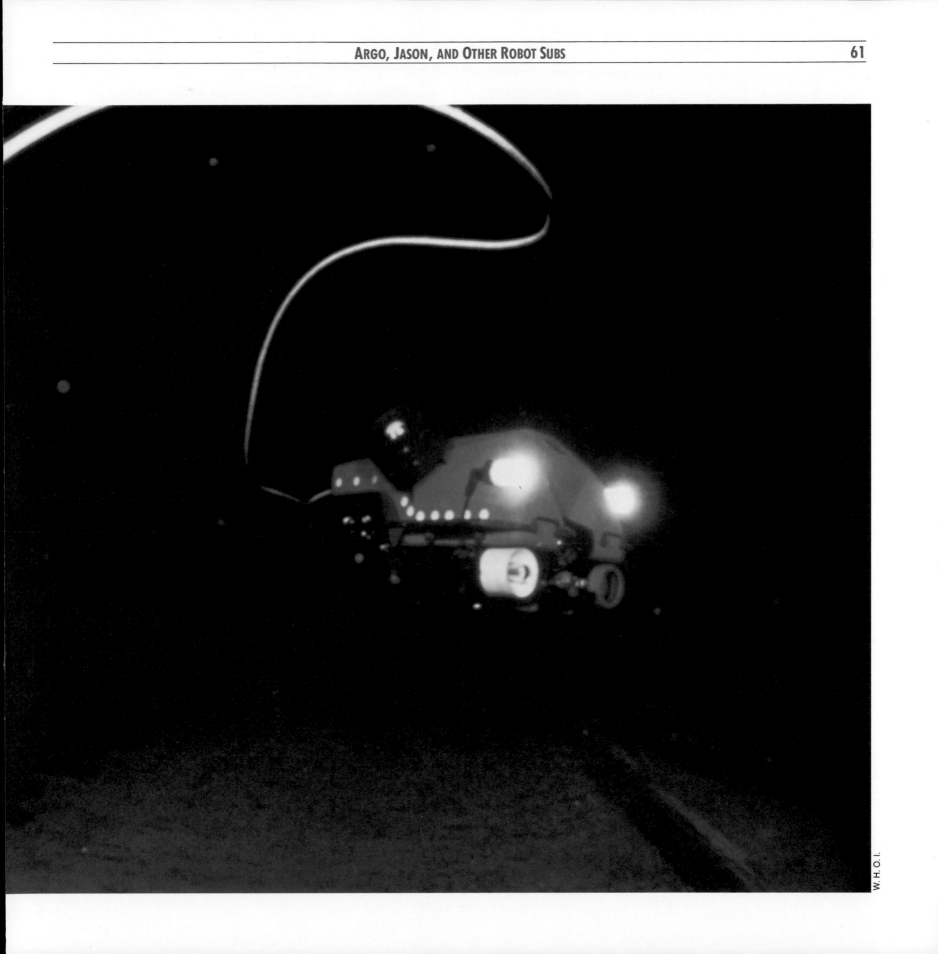

JJ—sending it into caves, under overhangs, or into other dangerous areas," Yoerger says. "But if that doesn't happen, parts of *JJ* will live on in *Jason*."

WHOI and other scientific institutions do not have a patent on unmanned submersibles. In fact, hundreds of small robot subs are currently in use off the coasts of the United States and elsewhere. Many are used by the United States Armed Forces, but the rest are operated for commercial purposes. They are known as ROVs— remotely operated vehicles—and are replacing human divers in many dangerous and exhausting underwater projects.

Commercial ROVs were introduced to the submersible market in the 1970s, as demand for risky work on subsurface drilling equip-

Today's commercial ROVs, like *Phantom 500* (right), replace humans in certain life-threatening tasks required to repair undersea drilling rigs, telephone lines, and other equipment.

ment grew. Even today, however, as more and more complex models are unveiled, most are far less ambitious than *Jason*. Operated by a tether leading to the surface, they are incapable of descending anywhere near the 20,000 feet (6,100 meters) WHOI's submersibles have reached.

Even so, the capabilities and accomplishments of commercial ROVs are very impressive. *Scarab*, for example, designed and built by California's Ametek Straza (one of the largest producers of ROVs) is equipped with two hydraulic manipulator arms, video cameras, depth and temperature sensors, and a wide range of tools. Under joint ownership of a group of communications companies, *Scarab*, weighing in at 3 tons (2.7 metric tons), is perfectly designed for such

The rusting structure of a drilling rig (left) presents an ominous sight that does not stop the agile, powerful, and hardy ROVs.

Courtesy Ametek Straza

underwater applications as locating, excavating, recovering, and replacing telephone cables at a depth of up to 6,000 feet (1,800 meters). *Scarab* also made news in 1985 when it tracked down signals from the flight recorder of a sabotaged Air India jet in even deeper waters (6,700 feet [2,050 meters]) off Ireland and then spotted and retrieved the vitally important plane recorder from the rubble of the ocean floor.

According to Jim Snow, marketing manager of Ametek Straza's ROV Systems, another of the company's subs has participated in an important mission. "When the space shuttle *Challenger* exploded, a whole flotilla of ROVs began to search for it," he recalls. "One of them was our *Gemini,* which helped recover the vital solid rocket booster."

Gemini was able to accomplish this delicate task because of its high-powered thrusters, powerful stabilizers, and agile hydraulic hands. Like other advanced commercial ROVs, it is designed to ac-

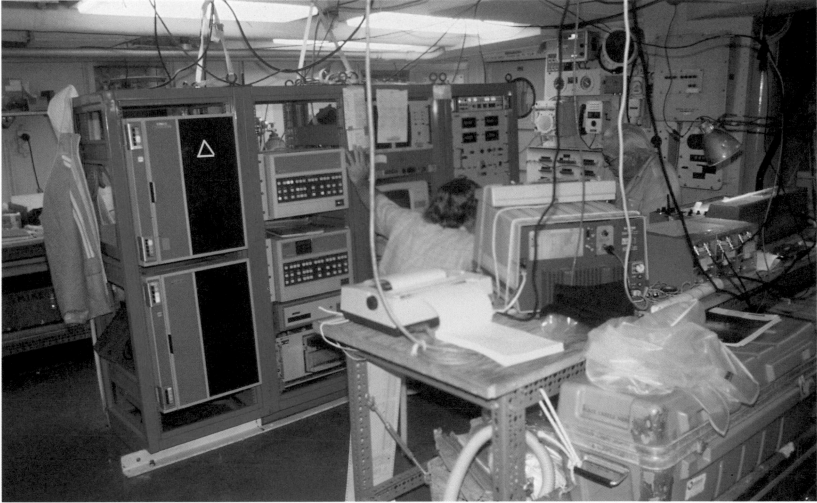

W. H. O. I./John Porteaus

cept instrument modules—self-enclosed optional packages that enable it to journey as deep as 10,000 feet (3,050 meters)—carry video tape recorders, sonar and other advanced instruments.

Ametek Straza is just one of the companies currently producing highly advanced commercial ROVs. Perry Offshore, a Florida company, recently participated in the search for the wreckage of the *Challenger* with its newest ROV, *Sprint 101*, while California's Hydro Products has also designed several ROVs, in-

cluding the *RCV-225*, a "swimming eyeball" with an 8-foot (2.4-meter) manipulator arm. Like many other ROVs, it can stay underwater for days at a time.

No matter how ambitious these vehicles are, however, they do not approach in complexity what some scientists believe is the wave of the future for submersibles. These are the coming generation of un-manned, untethered ROVs—subs that use powerful computers to undertake surveys, underwater repair work, and other challenging tasks.

Some ROV's (left) are "swimming eyeballs" that send data to scientists in surface laboratories (above), many of which look much like this control room for *ANGUS.*

The first untethered ROVs have already hit the water. Perhaps the earliest of all such ROVs in commission was the *Epaulard,* a French-built autonomous underwater vehicle (AUV) that has made hundreds of dives since its unveiling in 1979. The *Epaulard* is able to follow a programmed route, taking both still pictures and video images of the bottom before returning to the surface.

A far more advanced AUV—and the first one designed for mass production—is the *Autonomous Remote Controlled Submersible,* built by Canada's International Submarine Engineering. This remarkable sub, with a built-in computer, can dive to depths of 1,200 feet (360 meters), surveying the ocean bottom, monitoring the water's salinity, and gathering other data with delicate instruments—completely independent of operation from the surface. It will be particularly useful, its designers hope, for such difficult work as surveying the frigid ocean under the Arctic ice fields in search of likely oil-drilling sites.

Perhaps the most futuristic autonomous vehicle of all is still under development at the University of New Hampshire. This is the *Experimental Autonomous Vehicle East (EAVE-East),* which boasts a startling five microprocessors on board to aid it in its work inspecting pipelines and other undersea objects.

EAVE-East's computers record and analyze changes in depth, potential obstacles, and other data supplied by the craft's delicate sensors, which include sonar and pressure gauges. They also keep a

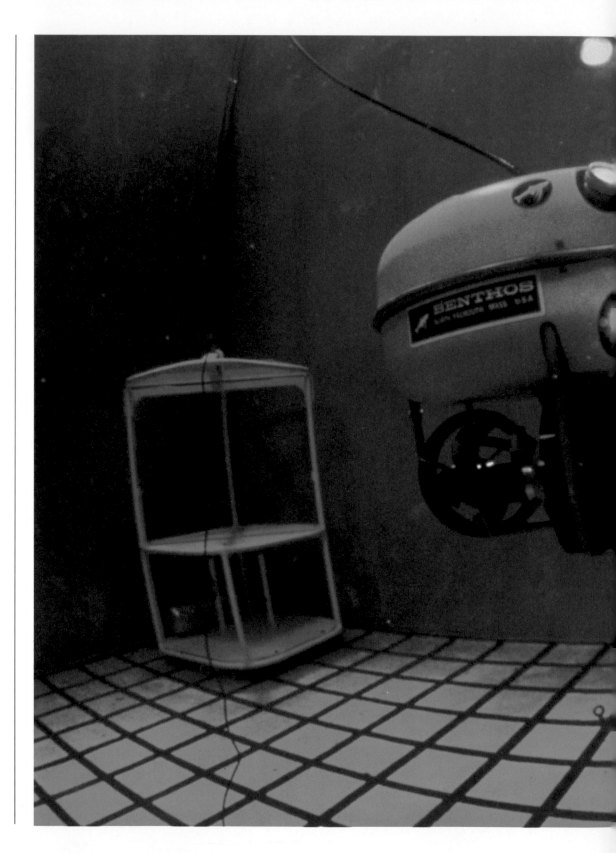

Lou Lehmann, courtesy Can-Dive Services Ltd.

W. H. O. I./Robert Ballard

Jason and other tethered subs look like vehicles of the future—but technological advances are moving toward models (left and above) that may someday operate without any human assistance.

W. H. O. I.

Dana Yoerger (left), of Woods Hole's Deep Submergence Laboratory (DSL), is working on a new generation of research subs—unmanned vehicles that can explore the deep-sea floor.

constant eye on the vehicle's power supply—and, using a programmed map of the terrain being surveyed—are even able to chart a new course to save power or avoid an unexpected obstacle.

Currently, *EAVE-East* is undergoing its second year of testing in Lake Winnipesaukee, New Hampshire, soon to be followed by ocean testing. And then, the sub's designers hope, *EAVE-East* will begin the most rigorous trials of all: tests under the sort of harsh natural conditions that, as WHOI's Yoerger says, "tell you more about your submersible than all the laboratory experiments and shallow-water tests put together."

Yoerger, who is still working to perfect the tethered *Jason,* is watching the development of *EAVE-East* with great interest. "I think its designers are doing fascinating work," he says, "but it and all the commercial ROVs have very little in common with *Argo/Jason.* They are designed to perform a variety of comparatively simple tasks, while our job is to devise a vehicle capable of exploring areas of the deep-sea that have never been visited before."

Although they may have varying purposes, *Jason, Jr., Scarab, RCV-225, EAVE-East,* and the hundreds of other unmanned submersibles are together exerting an unprecedented impact on our ability to gain knowledge and understanding of the vast, mysterious expanse of the deep sea. With their help, we are guaranteed to learn more about—and, increasingly, profit from—the ocean in the next few years than we have in the entire history of ocean exploration.

But where does all of this unmanned submersible development leave *Alvin, Deep Rover,* and other manned submersibles? As the untethered vehicles become more sophisticated, they'll be able to accomplish many of the tasks that are now the exclusive domain of the manned subs, while being less risky and less expensive. Do these advantages—which are bound to grow even more pronounced in years to come—mean that the manned sub is a dinosaur, destined for extinction?

Yoerger and Walden think not. Walden, who is in charge of keeping *Alvin* operating, says, "*Argo/Jason*'s work will be extremely important; I'm eager to see what they find, but I firmly believe that there will always be a need for submersibles that allow scientists to go down and actually look at what they're studying."

Yoerger concurs. "Although I believe that *Jason*'s abilities will overlap with *Alvin*'s, I also think that you simply cannot discount the 'eyeball effect'—the excitement that researchers get when they've been down below on *Alvin.*"

A perfect example of this, Yoerger says, was the exploration of the *Titanic* in 1986. "Sure, the pictures that *Argo* sent back earlier were exciting, but no one was satisfied with that," he recalls. "They wanted to visit the wreck themselves."

*E*AVE-East (left), currently being tested, uses computer technology to avoid obstacles and perform complicated tasks without human intervention.

TREASURE HUNTS

"The main reason we chose to send *Argo* to look for the *Titanic*," WHOI's Yoerger explains, "is because the search area was close to home, and the terrain was benign. It seemed an easy, appropriate way to test out the sub's capabilities."

Other Deep Submergence Lab scientists agree that the choice of locale for *Argo*'s maiden voyage was based on logical considerations—what was best for the sub, its operators, and its designers. Without doubt, these were truly their priorities.

But excitement of these same scientists in 1985, when *Argo* found the long-lost luxury liner in the frigid waters of the North Atlantic, revealed another, perhaps more compelling truth: The discovery and exploration of sunken ships, from battleships lost at war to Spanish galleons laden with treasure, seems to hold a special fascination for researchers and the public alike. Today, robot subs, advanced surveying techniques, and professional treasure hunters are making an enthusiastic new effort to find and salvage sunken ships. In doing so, they have enriched our knowledge of both the recent past and ancient or legendary civilizations.

No search has been more representative of the marriage of science and adventure than the hunt for the *Titanic*. As is well known, the 900-foot (275-meter) liner set out on its transatlantic maiden voyage in 1912 as the grandest, most luxuri-

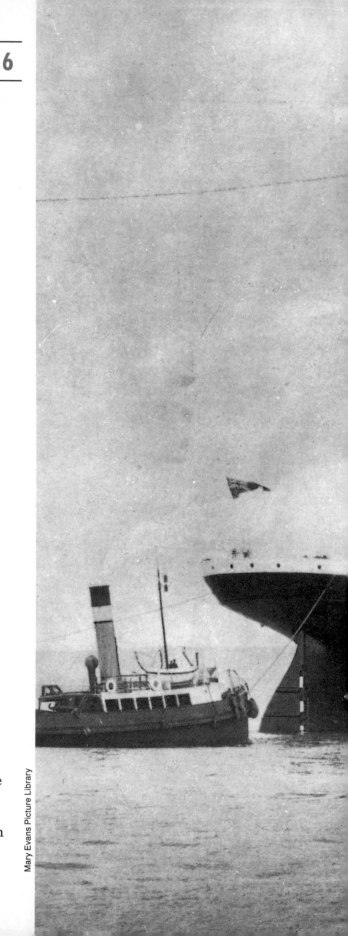

Since its disastrous demise in 1912, the great luxury liner *Titanic* (previous page) has become the world's most famous—and most sought-after—sunken ship.

WHOI's research vessel, *Knorr* (far right), carrying the submersible *Argo* (right), steamed to the North Atlantic in 1985 to join the hunt for the *Titanic*. The *Atlantis II*, transporting *Alvin*, followed soon after.

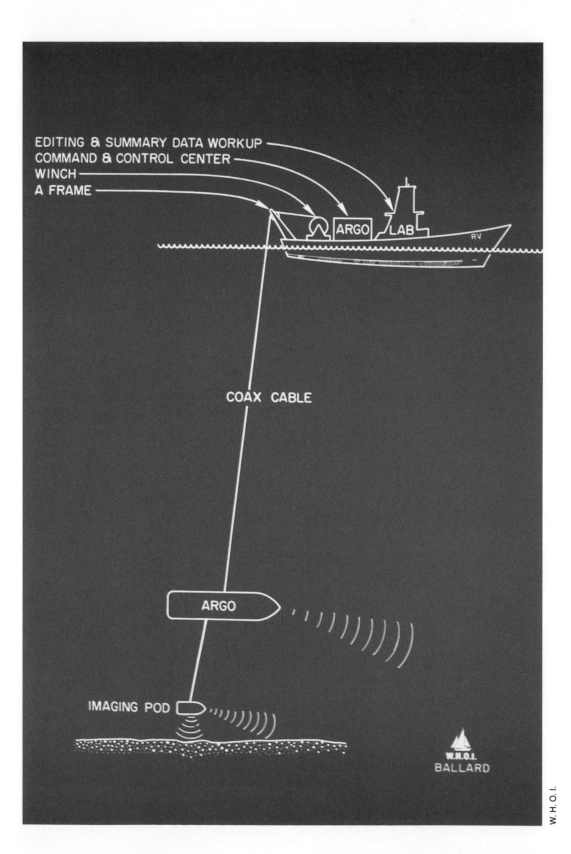

EDITING & SUMMARY DATA WORKUP
COMMAND & CONTROL CENTER
WINCH
A FRAME

ARGO LAB

RV

COAX CABLE

ARGO

IMAGING POD

W.H.O.I.
BALLARD

W. H. O. I./Anne I. Rabushka

ous ship ever built. It was considered indestructible, the finest achievement of an increasingly technological age.

Then, one dark night, the impossible happened: The *Titanic* struck an iceberg and quickly sank. While a nearby ship failed to heed the sinking giant's distress calls, passengers aboard the great liner discovered that the "indestructible" ship was not equipped with enough lifeboats to carry everyone to safety. As a result, more than 1,500 people died in the worst shipwreck in history.

In the decades following the *Titanic* tragedy, the ship's legend has continuously grown. Books, movies, songs—all have endlessly rehashed the events leading up to that tragic night. But, until recently, no one dreamed of going down to the bottom of the North Atlantic and actually *finding* the remains of the famous ocean liner. The technology simply did not exist to make such a search possible.

Then, beginning in 1980, with the technology in hand, teams of explorers used sonar to survey the area where the ship was thought to have sunk. Jack Grimm, a Texas millionaire, sponsored three separate searches. During one of these attempts, expedition leaders reported having possibly found the wreck, but bad weather intervened. As a result, that search was eventually dropped, as all others had been, without any confirmation that the *Titanic* had actually been found.

Finally, in 1985 the first rigorous, truly scientific hunt was attempted. At that time, WHOI research ship the *Knorr,* outfitted to tow the *Argo,* joined with a French team of scientists aboard the ship *Le Suroit,* which carried the submersible *Sar.* Like *Argo, Sar* was equipped with a sonar device to provide high-resolution "sound maps" of the ocean floor, as well as other complex surveying devices. Unlike WHOI's sub, *Sar* did not contain advanced video systems for a view of the floor.

Sar had the first crack at finding the wreck. For more than three weeks, during the summer of 1985, it swept across the 150-square-mile (390-square-kilometer) area where researchers thought the *Titanic* went down. The submersible found nothing, but, as Yoerger points out, in coming up empty the French researchers performed a valuable task. "A large part of deep-sea sur-

veying is pretty uneventful," he says. "But, at worst, it tells us where not to look again."

Next, in August, came the *Knorr*'s and *Argo*'s turn. They were operated under the direction of Ballard, the head of the Deep Submergence Lab, and *Argo*'s designer. Towed at 12,500 feet (3,800 meters) below the surface, the sub scanned the ocean bottom with both video cameras and sonar. At first, it came up with nothing—but on September 1, the new sub (and its operators) struck pay dirt.

Aboard the *Knorr,* scientists watching the video screens suddenly saw *Argo* glide past a hulking metal object. It was, they knew almost immediately, one of the *Titanic*'s vast boilers. Within a few minutes, *Argo*'s cameras located more of the sunken liner, and the

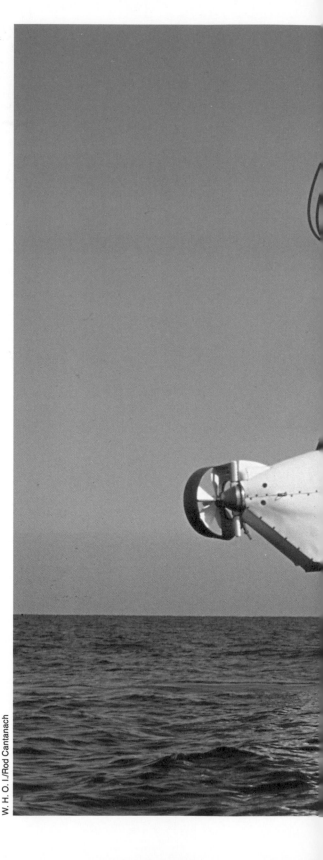

Argo was the first sub to spot the *Titanic. Alvin,* pictured here suspended from *Atlantis II,* however, allowed scientists to actually explore the sunken liner.

W. H. O. I./Rod Cantanach

The robot sub, *Jason, Jr.,* with a network of strobe lights and cameras, actually ventured inside the ravaged hulk of the *Titanic.*

searchers knew they had finally found the final resting place of the great ship.

Argo, along with *ANGUS,* WHOI's other towed camera vehicle, spent four more days surveying the wreckage, but the next truly dramatic sightings of the *Titanic* didn't occur until a subsequent mission nine months later. These remarkable searches had to wait until July 1986, when Ballard and a team of scientists returned to the wreck site aboard the *Atlantis II.* As usual, the research vessel was carrying the *Alvin,* the manned sub that has played such a large role in so many marine adventures.

During the following two weeks, researchers aboard the sub took eleven long journeys 12,500 (3,800 meters) feet straight down to view the remains of the *Titanic.* Along with *Alvin*'s usual complement of gauges and instruments was the robot sub *JJ,* with a tether long enough to allow it to venture inside the great rusting hulk.

What the explorers saw and filmed (they returned with hours of videotape, while *ANGUS* took tens of thousands of photographs), was a dramatic testimony to both the destructive and the preservative forces of the deep sea. The *Titanic* apparently broke into two, either while still on the surface or while it sank, because its bow and stern are now separated by nearly 2,000 feet (610 meters).

None of the ship's woodwork has survived; all has been eaten by voracious worms. The ship's brass fixtures and other metal objects, however, remain in pristine condition, buffed to a bright shine by the passing currents.

Other artifacts, on a smaller, more personal, and more poignant scale, were a sobering testimony to the disaster. Atop a huge boiler, for example, the scientists spotted an unbroken porcelain coffee cup. It must have floated down in the wake of the *Titanic*'s tragic voyage to the bottom, finally sitting up-

right as if just recently placed there. Intact wine bottles were scattered amid the rubble that lay behind the ship's upright bow, while among thousands of artifacts searchers also saw a patent-leather shoe, the porcelain head of a doll, and several small safes.

The most thrilling views of all were recorded by *JJ,* on these, its first four voyages. Despite its occasional motor problems, *JJ* was able to snake its way down through four of the *Titanic*'s decks. The little robot entered ballrooms and a gymnasium, filming elegant light fixtures and other preserved objects, including a brass plaque that read, "This door for crew use only." Neither *JJ* nor any of the other searchers spotted any human remains, all of which must have long since disintegrated.

Ballard made one discovery that seems certain to keep the debate over how the *Titanic* sank alive indefinitely. For years, most experts have believed that the great liner's

Today's ocean researchers use advanced instruments, such as this WHOI subsurface buoy (right), to map the sea floor, measure water temperatures, and perform other exploratory tasks.

W. H. O. I./George Hampson

collision with an iceberg resulted in a huge gash in the starboard side of its hull. This gash, of course, would then have been responsible for the ship's sinking.

But the explorers, in their detailed survey of the hulk, found no evidence of any gash. Instead, Ballard believes, the iceberg's impact caused the steel plates of the *Titanic*'s hull to buckle inward. Some of the rivets holding the plates together might then have popped, allowing water to rush in. This would have caused the gap between the plates to widen further, forcing more rivets to pop, and allowing more water to pour in.

While the *Titanic* is by far the most famous recently explored shipwreck, it is certainly not the oldest or the most revealing. In recent years, armed with an array of powerful new surveying devices, scientists and other adventurers have begun to discover increasing numbers of wrecks—many hundreds, or even thousands, of years old—that are helping reshape our vision of the past.

Explorations are currently underway nearly everywhere throughout the world, as marine archaeologists and treasure hunters alike seek to find ships long thought lost forever. Almost always, the searchers depend on three recent advances in surveying technology: side-scan sonar, the magnetometer, and the sub-bottom profiler.

Side-scan sonar is the acoustic mapping device found on board the *Argo* and many other new submersibles, but it can also be used by surface vessels. It consists of a 4-foot (1.2 meter) metal "fish"—complete with stabilizing fins—towed through the water behind a ship. The fish emits high-frequency pulses of sound that are sent both downward and sideways in a fan-shaped beam that scans the seabed below, encompassing far more of the sea floor than older sonars. (This ability to scan sideways as well as straight down earns the device its name.)

When the sound waves transmitted by the metal fish bounce off the sea floor, some return to a receiver on the device. The results are then translated into visual images printed on paper in a rotating drum. The darkness of the final image depends on both the distance the sound waves travel and the density of the object they bounce from. Hills, valleys, rocks, mud, and sand

Scientists in the *Atlantis II*'s video room (left) were the first to spot the *Titanic,* as filmed by *Argo* in the cold waters of the North Atlantic. Side-scan sonar allows researchers to map huge areas of the mysterious ocean floor.

all produce their own quickly recognizable images. The final result is a detailed map of the bottom, nearly as informative as a photograph, but without the need of strobe lights and film.

In the search for wrecks, side-scan sonar is both an important and a limited tool. The instrument is capable of mapping the surface of the sea floor in great detail. In detecting a wreck lying on the bottom, it may provide an image detailed enough that the ship can actually be identified before divers go down to investigate it. But some wrecks are hundreds or thousands of years old and may well have suffered the ravages of a harsh resting place—including being buried in sand or silt, beyond the view of sonar. That's where the sub-bottom profiler comes in handy.

The profiler, like side-scan sonar, uses sound waves to map the sea floor. But the pulses it sends downward, in a conical beam, are of low instead of high frequency, and low frequency sound waves are able to penetrate many layers of sediment. The changes between different layers (for example, where sand rests on silt, or silt on bedrock) are also recorded, making the profiler a valuable tool for marine geologists. But for those hunting shipwrecks, the profiler's ability to spot ships buried under many feet of sand makes the device so exciting.

Even more remarkable is the magnetometer, the third important search tool, which actually uses the earth's own magnetic field as a guide in the hunt for sunken ships. The earth's magnetic field is generally stable and predictable in any given area and sends out a constant signal, which can be picked up by a small measuring device, a bottle, that is towed through the water by a search vessel.

Certain conditions, however, can interfere with the magnetic field's natural signal, most significantly—and of greatest interest to treasure seekers—concentrations of such metals as iron or steel. Reinforced hulls, ship's fittings, iron cannons, and other objects can all cause a change in the signal. The bottle registers these magnetic variations, whose size and shape also provide information on the size, depth beneath the surface, and possible depth of burial under the ocean floor of the metal objects down below. These variations are relayed along a cable to researchers on a ship located on the surface;

W.H.O.I.

After lying undiscovered for more than seven decades, the hulk of the *Titanic* (above and right) presented an eerie sight to researchers aboard *Alvin*.

The ancient coastal city of Caesarea (left), built more than 2,000 years ago, was rediscovered recently by Israeli scientists. Today, a restoration effort is underway.

Under the placid waters of this sunlit bay (right) lie the ruins of ancient Caesarea, once a bustling, sophisticated port city ruled by King Herod.

Courtesy of Israel Government Tourist Office

the researchers then know they've found something unusual.

Today's most efficient magnetometers are able to spot a large steel wreck at depths of up to 600 feet (180 meters) and an individual iron cannon at 100 feet (30 meters). But they also have drawbacks. Since bronze objects don't interfere with the magnetic field, they don't show up at all. And smaller iron or steel objects will only register if the bottle is brought quite close to them. In cases of deeply buried wrecks, however, and those in areas that

Courtesy of Israel Government Tourist Office

are hard to scan with sonar, the magnetometer is an important search device.

As a result of the tracking capabilities made possible by side-scan sonar, profilers, and magnetometers, the past few years have seen a surge of interest in discovering and investigating shipwrecks. In England, for example, the Marine Archaeological Survey was formed in 1985 to survey nearby waters, providing the first comprehensive register of Britain's sunken ships. And the United Nations is spearhead-

ing an ambitious survey of the world's undersea treasures.

Other countries have already devoted years to searching for the wrecks that lie in their waters. The Undersea Exploration Society of Israel (UESI), for example, was created in 1958, when fishermen in a kibbutz decided they should develop a better way of recovering the ancient amphoras and other relics they kept dragging up in their nets. The fishermen called in a group of divers from the Israeli Navy, and these divers became in-

terested in marine archaeology. As a result, a nationwide effort to investigate the region's past was born.

In recent years, UESI and other groups have uncovered evidence of several elaborate ancient harbors, but none more impressive than Caesarea, built under the orders of King Herod over a twelve-year period beginning in 22 B.C. The excavation of this vast harbor, complete with sophisticated breakwaters and sunken ships (all now buried well underwater), was begun more

Archaeologists and other scientists have spent years recovering and restoring the wreck of an ancient Roman sailing vessel (right)—the so-called "Jesus boat"—that sank off the shores of Galilee 2,000 years ago (far right).

than a decade ago and continues to require the efforts of many divers today.

In 1986, international attention was focused on another Israeli find. Most of the scientists involved agree that it is an important historical discovery: a two-thousand-year-old wooden boat resting on the bottom of the Sea of Galilee, miraculously well preserved because it was covered in a layer of thick silt. Among the artifacts found in the area are items dating both from before and after the first century A.D.

What makes the discovery so special to some people is its possible religious significance. The Bible says that Jesus crossed the Sea of Galilee in a fishing boat after walking on the water. This fishing boat dates from approximately the right time. Religious pilgrims have, therefore, been hurrying to

Courtesy of Israel Government Tourist Office

Courtesy of Israel Government Tourist Office

the site, hoping for a glimpse of "Jesus' boat."

The chances that this particular boat is that once used by Christ are questionable at best. But to the researchers on the scene, the boat's scientific and archaeological value is inestimable, or will be if the fragile, extremely delicate hull can be preserved without falling into pieces.

The task of raising the wreck required extreme care—and a little luck. First, scientists covered the hull in polyurethane plastic (much like a home owner might cover a sofa). This plastic served as a buoyant bubble in which the boat could be gently floated to the surface and then carried intact to shore.

Having brought the boat safely to land, experts faced an even more difficult task: keeping it from disintegrating when exposed to air after centuries underwater. Their temporary solution: immersing the hull in a pool of water—which, in effect, duplicated the boat's ancient resting place offshore.

The boat is now in the midst of a painstakingly long process of preservation. The entire hull is currently being coated in polyethylene glycol, a type of wax. Eventually, this wax will saturate the wooden hull, replacing the water that has filled it for so long. When the process is complete, the boat will actually have been embalmed—and it will also be saved from any further deterioration.

Israel does not have a monopoly on the recovery of ancient boats. Among others discovered recently, the ship with perhaps the greatest historical value may be a wreck that wasn't even found through the use of modern technology. Off the

Courtesy of Israel Government Tourist Office

"Jesus' boat" was found—after more than two millennia of lying undisturbed—by divers searching these quiet Israeli waters in the Sea of Galilee (far left). Who knows what other fascinating treasures lie hidden there?

While scientists used sonar, computerized subs, and other advanced technology to locate the *Titanic*, Israel's more simply equipped divers have had great success locating sunken ships and other reminders of the country's rich history (left).

coast of Turkey, a team of local scientists and Americans from the Institute of Nautical Archaeology at Texas A&M University have been recovering the cargo of a wooden sailing ship that sank an amazing 3,400 years ago.

For thousands of years the broken-up ship rested in water ranging from 140 to 170 feet (42 to 52 meters) deep. Then, in 1982, a local sponge diver stumbled upon the wreck, and the fascinating work of recovering and studying its ancient cargo began. Thus far, divers have brought to the surface gold jewelry, pottery shards, small gold sculptures, weapons, and many pottery jugs containing trading goods. Researchers and historians have not yet revealed their opinion of the ancient ship's ori-

gins, but the discovery has already enlarged our understanding of the lives of the region's inhabitants thousands of years ago.

Sunken ships can also be found off the coasts of the United States, though none of these is quite so old. In fact, many of the most interesting wrecks are comparatively young, as befits a country that is itself little more than two hundred years old. However, some recent finds have given us important new insights into the brief but colorful history of the United States.

When we think of shipwrecks, we envision galleons breaking apart on hidden coral reefs, warships sinking in the stinging waves of the North Atlantic, or Renaissance-era sailing ships scuttled by pirates. We don't think of impor-

tant wrecks lying on the bottom of lakes. Yet Lake Champlain—a large lake shared by New York, Vermont and Quebec—may hold the last remnants of one of the most crucial battles in the Revolutionary War or the War of American Independence. It was here that, in his day as a patriot, Benedict Arnold, America's renowned traitor, led a tiny fleet of warships against powerful British forces in a successful effort to stem a planned invasion from Canada.

Many of the vessels of Arnold's 15-ship fleet were sunk during the battle. However, the gunboats, cutters, schooners, and other ships that comprised his navy were never considered "lost" at the bottom of Lake Champlain. As early as 1776, cannons and other objects were be-

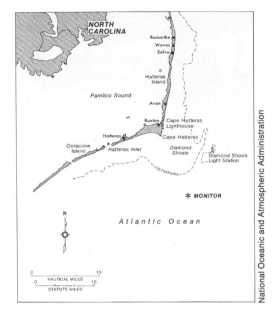

National Oceanic and Atmospheric Administration

ing recovered from several of the sunken ships, and as the decades passed, more of the wrecks were raised and then frequently either deteriorated or were stripped by relic hunters. Only one ship, the *Philadelphia,* is currently on display (at the Smithsonian Institution's Museum of History and Technology in Washington, D.C.)—and that was saved from destruction more through luck than design, a quarter century after it was first recovered.

Reports from the time of the battle are either absent or confusing, but experts believe that at least one of Arnold's ships still rests on Lake Champlain's muddy floor. The Champlain Maritime Society, using divers and side-scan sonar, has been active in searching for

relics of this important part of United States history. Unfortunately, this effort, valuable as it is, comes too late to do much more than pick up the pieces of this fascinating and important battle.

The most famous ship from another American war is also the object of a recent preservation effort. Although the Civil War ironclad warship *Monitor* sank in a storm in 1862, only in 1973 was it found in the Atlantic Ocean off the coast of North Carolina. In 1975, the site was designated the first National Marine Sanctuary in the United States.

The *Monitor,* a northern ship, was a participant in a pair of famous battles between the navies of the North and South during 1862, early in the Civil War. The *Moni-*

Resting off the coast of North Carolina (far left), the famous ironclad warship *USS Monitor* (left, on the day of its demise) wasn't found until 1973. Today, researchers hope to raise and restore the fragile wreck.

National Oceanic and Atmospheric Administration

tor's two battles, both against the southern *Merrimack,* were generally considered standoffs. Interestingly, historians think these skirmishes may have been indecisive because neither side wanted to risk losing a prized warship. But the importance of the *Monitor* far transcended its role in the Civil War. The first warship designed and built with iron armor, it was the forerunner of all battleships to come, sounding the death knell for the wooden ships that had dominated naval warfare for centuries.

The 1973 discovery of the wreck of the *Monitor*—lying upside down in 220 feet (67 meters) of water—prompted bitter arguments over what to do with the historic ship. Should it be salvaged or left where it was? If recovery was warranted, how could the ship be raised without damaging its corroded, but largely intact hull? Was the astronomical cost of recovery worth the benefit of saving the historic ship? Naming the site a Marine Sanctuary merely protected the old ironclad from plunderers; it did not settle any of these or other questions about the ship's future.

Finally, in 1985, the United States' National Trust for Historic Preservation and the National Oceanic and Atmospheric Administration (NOAA) announced a cooperative effort aimed at saving the *Monitor.* A large-scale recovery effort, including extensive documentation and surveying of the site, careful salvage efforts, and preparations for the preservation and display of the ship, is now underway. Whether the fragile hull will survive remains to be seen.

While the recovery of the *Monitor* and any remaining ships from Benedict Arnold's fleet is fascinating for historical reasons, the excitement of these finds cannot compare with that surrounding another discovery in 1985, in United States waters: a centuries-old Spanish galleon found in the Gulf of Mexico off the Florida Keys. Since its discovery, the *Nuestra Senora de Atocha,* which sank with its sister ships in a 1622 storm, has spurred much media attention.

What is the reason for all this excitement? Treasure—a fortune in gold, silver, emeralds, and other rare items, lay buried in this ship on the ocean floor for more than three hundred years until a professional treasure hunter named Mel Fisher went to find them.

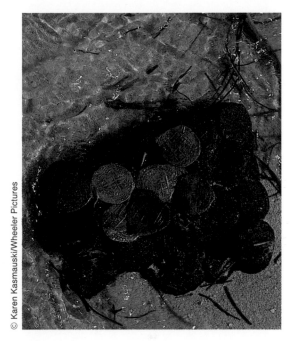

Treasures beyond imagination: silver, gold, and gems were only part of the riches recovered from the Spanish galleon *Atocha* (left and above).

Fisher didn't just stumble upon the wreck of the *Atocha*. As early as 1960, he had heard about the wreck's existence somewhere off the southern coast of North America. Within a few years, after researching the subject, he knew for sure that the galleon had gone down somewhere near the Florida Keys. Though this information focused Fisher's search somewhat, it still left vast areas of open ocean to search.

Over the course of fifteen years beginning in 1970, Fisher and members of his professional treasure-hunting company, Treasure Salvors, used sonar to map the sea floor and magnetometers to search for telltale magnetic variations caused by the ship's iron an-chors or other objects. In 1973, the searchers found a silver bar that they proved came from the *Atocha*, but twelve more years passed before they came upon the main site. Buried in the silty ocean floor was treasure beyond Fisher's wildest imaginings—or those of the 1,140 people who made serious monetary investments in the quest.

By late 1986, Treasure Salvors had brought to the surface more than 100,000 silver coins; 1,000 silver bars, weighing about 75 pounds (34 kilograms) each; nearly 3,000 emeralds; 150 gold bars, weighing up to 77 troy ounces (217 grams); several gold coins and chains; and many other artifacts. According to the company, the present bounty may represent only about half the treasure to be found at the site.

What does this mean for the lucky investors? Each will probably receive eighteen times the initial investment—a handsome return in such an uncertain business. The value of the *Atocha*'s treasure is uncertain, but it may approach 200 million American dollars.

The publicity generated by the *Atocha* haul has provoked a gold rush to find other treasure-laden wrecks. For many serious historians and marine archaeologists, this is a troubling development. While Treasure Salvors boasts an archaeological director whose task it is to record important information about the *Atocha* and its treasure, other treasure hunters are less careful. Centuries' worth of

Mel Fisher, today a happy and wealthy man, spent more than twenty years searching for the *Atocha*. Not until 1985 was his quest rewarded with amazing riches (left and right).

© Karen Kasmauski/Wheeler Pictures

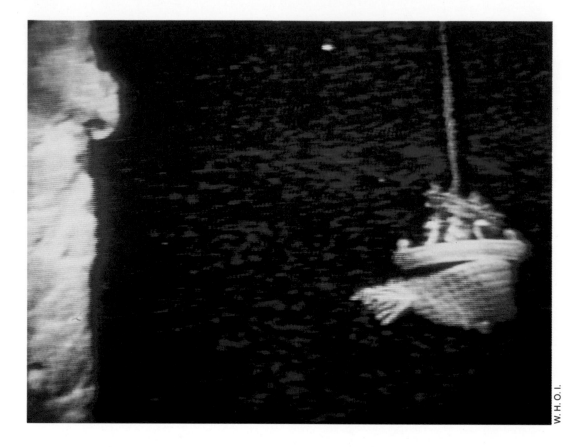

W.H.O.I.

While many scientists—including Robert Ballard—believe that the *Titanic* (left and right) should remain untouched, others are planning to bring as much of the ship as possible back to the surface.

fascinating information can be destroyed in just a few hours of mindless salvaging, much as the famous Mayan and Incan ruins of Latin America have been plundered for a booming illegal trade in artifacts.

Even the *Titanic* may not be safe. While Robert Ballard and the other explorers of the famous liner believe strongly that it should be left where and as it is, rumblings of another kind are already being heard. Grimm, the Texas millionaire who sponsored several fruitless efforts to find the ship, began planning a submersible hunt for relics almost as soon as word of the *Titanic* discovery reached the public. He was also part of a planned live television broadcast, scheduled for the summer of 1987, in which three

of the ship's safes would be opened, presumably before an audience of millions. And, in late 1986, President Reagan's staff began drawing up a program for the protection of the ship, the initial drafts of which reportedly include plans to raise parts of the wreck.

"We see *Argo* and *Jason* as being used solely for scientific and technological goals, not as commercial vehicles," says WHOI's Yoerger. Those priorities, however, may not be shared by the designers and operators of other submersibles and advanced search devices. Individuals may become far richer because of the fruits of modern treasure-hunting technology—but it would be tragic if our knowledge of the world and its history suffered.

CHAPTER 7

OCEAN DRILLING AND EL NIÑO

During the past decade, both manned and unmanned submersibles have opened our eyes to the fascination of the deep sea. With their strobe lights and video cameras, they have carried us among the volcanoes of the Mid-Ocean Ridge, brought us eye-to-eye with creatures living in the poisonous waters of the smoking vents, and taken us along on a jaunt up the grand staircases of the *Titanic*.

Yet all these achievements have revealed only a small fraction of what there is to learn about the undersea world. And submersibles—even the most futuristic of them—can't help us uncover some of the ocean's secrets. For example, some of the most exciting clues remain hidden from view because they lie under the ocean floor itself.

These clues are the keys to our planet's history as written in its crust: tales of ancient climates, the ebb and flow of primeval seas, the creation of mountains and valleys. Frustratingly, these insights have long been as inaccessible as the distant stars—but modern technology has provided a way to read them. It's a technique called ocean drilling, a method by which we can gather samples of the earth's core for intensive scientific analysis.

The first project to bring undersea rock to the surface was the Deep Sea Drilling Program (DSDP), sponsored beginning in 1968 by the Scripps Institution of Oceanography in La Jolla, California. For fifteen years, the *Glomar*

W. H. O. I./James Broda

Scientists throughout the world are studying the remarkable—and poorly understood—relationship between the oceans' and the earth's climate (previous page).

The Ocean Drilling Program (ODP) began its 10-year voyage to the deep-sea crust in 1985. Its mission: to study the distant past by exploring the earth's core.

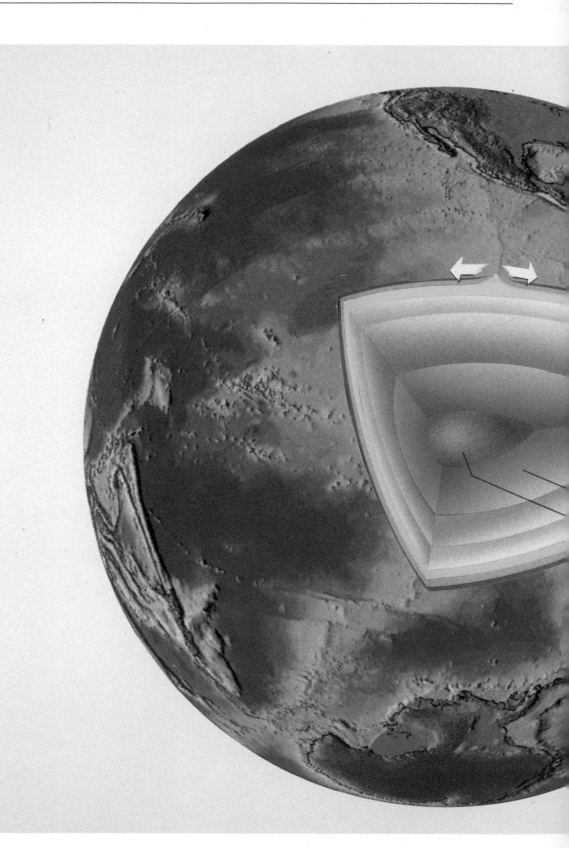

W. H. O. I./E. Paul Oberlander

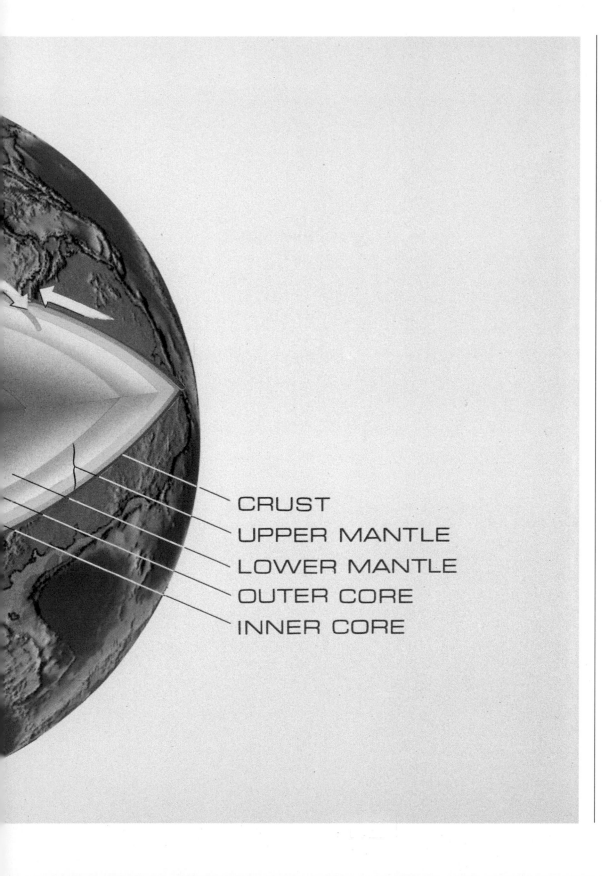

CRUST
UPPER MANTLE
LOWER MANTLE
OUTER CORE
INNER CORE

Challenger, a research vessel designed to spend months at a time out at sea, sailed around the world using a massive, hollow drill pipe, tipped with powerful diamond drill bits, to dig more than 1,000 holes in the sea floor. By digging down through the rock and bringing up core samples—carefully preserved rock specimens—researchers were able to gain the first insights into the structure and composition of the earth's deep-sea crust.

Though the DSDP ended in 1983, it was soon followed by the Ocean Drilling Program (ODP), a far more ambitious international effort, which began in 1985 and is slated to continue for ten years. The program, operated by Texas A&M University, is funded by the National Science Foundation, Canada, West Germany, France, Japan, and Britain. Planning and advice are provided by the Joint Oceanographic Institutions for Deep Earth Sampling (JOIDES), an international group of scientists.

The program's first goal was to find a new, more powerful drill ship to replace the *Glomar Challenger.* "We were looking for a large, sturdy research vessel, capable of functioning in extremely rough seas while accommodating dozens of scientists and state-of-the-art laboratories," recalls Professor Philip Rabinowitz, director of the ODP and a professor of oceanography at Texas A&M University. The ship had to house labs spacious enough to accommodate powerful electron microscopes and other advanced equipment, while also having room for huge quantities of drilling equipment.

Courtesy Ocean Drilling Program, Texas A & M University

Courtesy Deep Sea Drilling Project, Scripps Institution of Oceanography

The ship Rabinowitz and the others eventually found surpassed their expectations. It is a 470-foot (143-meter) vessel, previously used for commercial drilling, and is equipped with a powerful positioning system (a set of computer driven engines that automatically keep the ship stable while drilling). It has the capacity to house and feed more than one hundred scientists for seventy continuous days. ODP officials soon had the ship strengthened, outfitted with seven floors of laboratories, and modified to accommodate a drill 30,000 feet (9,144 meters) long. The ship's new name: the *JOIDES Resolution.*

The newly outfitted vessel took to the water in January 1985. The first journey, dubbed Leg 100, was largely a trial run for the rebuilt ship, aimed at testing such features as the dynamic positioning system under harsh conditions. The *Resolution*'s operators found that the ship functioned perfectly during the eighteen-day test, even when exposed to 50-knot winds and 20-foot (6-meter) waves.

Despite the success of this battery of tests, Leg 101, the first actual research cruise, required no such sturdiness from the *Resolution*: Beginning in February 1985, it carried a scientific team to the Bahamas, where the ship flawlessly retrieved core samples from the world's largest known deposit of calcium carbonate (the decomposed remains of ancient shelled creatures). In June of the same year, the *Resolution* traveled to the waters off Spain. Research from this trip helped scientists understand the process by which the earth's crust tore apart in that area, separating the continent of North America from Africa and Europe, more than one hundred million years ago.

These research trips, like all those that followed, were planned similarly, says Rabinowitz. "Once we had a firm idea of the goals of a specific cruise, we would send out information asking for applicants from all over the world," he explains. "Out of the huge list of applicants, we'd then choose those that fit best with our goals—at least two people from every member country." Two chief scientists would go on each trip, with at least one Texas A&M staff scientist.

Perhaps the most exciting—and

A remarkable amount of advanced technology is put to use on each leg of the Ocean Drilling Program (ODP) (left and right). Sonar, thrusters, and positioning systems all must work perfectly for a successful mission to take place.

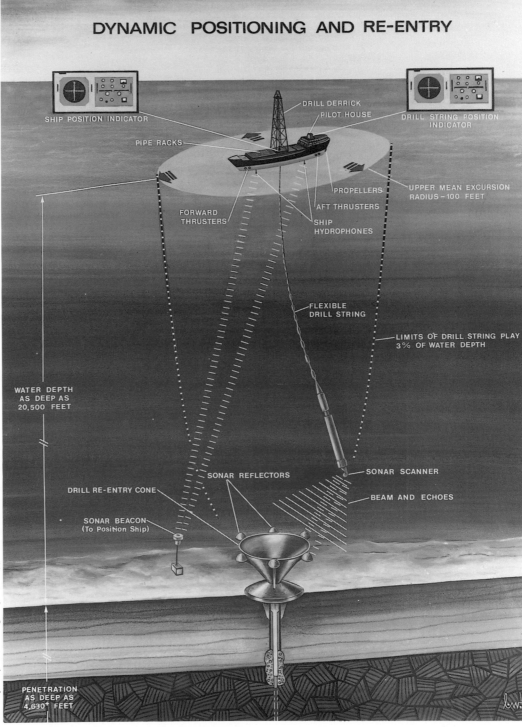

DYNAMIC POSITIONING AND RE-ENTRY

SHIP POSITION INDICATOR

DRILL STRING POSITION INDICATOR

DRILL DERRICK
PILOT HOUSE

PIPE RACKS

PROPELLERS

UPPER MEAN EXCURSION
RADIUS ~ 100 FEET

FORWARD
THRUSTERS

AFT THRUSTERS

SHIP
HYDROPHONES

FLEXIBLE
DRILL STRING

LIMITS OF DRILL STRING PLAY
3% OF WATER DEPTH

WATER DEPTH
AS DEEP AS
20,500 FEET

SONAR REFLECTORS

SONAR SCANNER

DRILL RE-ENTRY CONE

BEAM AND ECHOES

SONAR BEACON
(To Position Ship)

PENETRATION
AS DEEP AS
4,630+ FEET

Courtesy Deep Sea Drilling Project, Scripps Institution of Oceanography

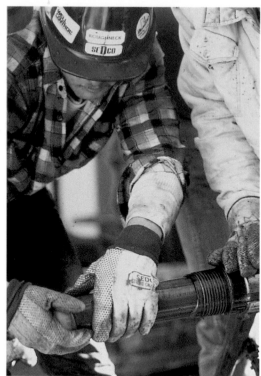

Courtesy Ocean Drilling Program, Texas A & M University

Heinz Steenmans/Wheeler Pictures

During the ODP's jaunt to Baffin Bay in 1985, icebergs presented a constant threat (left). The research, however, was worth the harsh conditions. Every part of the project was technologically complex (far left and right).

Courtesy Ocean Drilling Program, Texas A & M University

certainly the most dangerous—of the ODP's first ten missions was Leg 105, which took the ship to Baffin Bay off the west coast of Greenland in September and October of 1985. Never before had a drill ship ventured to such high latitudes, yet the *Resolution* braved icebergs and howling Arctic storms (unavoidable, even with careful planning, as they can occur at any time of the year) to take important core samples from 3,500 feet (1,066 meters) below the ocean floor.

"On this trip, we were accompanied by an ice-patrol vessel," Rabinowitz says. "Although we proved that the *Resolution* is well equipped to handle even the harshest conditions, when you're surrounded by icebergs, someone had better keep an eye on them."

This leg, as well as several others, delved into the mysteries of the Baffin Bay area's ancient weather and how that climate may have influenced the development of life on earth. The geographical characteristics of the area played an important part in determining that climate, by providing a corridor for the exchange of water between the frigid Arctic Ocean and the warmer Atlantic Ocean more than 60 million years ago. This mingling of oceans is thought to have had a major long-term effect on region's (and the world's) weather, causing long-term cooling in some areas and warming trends in others. By microscopic study of the earth's core, with attention to the differences in the composition of rock and the presence or absence of specific living organisms, researchers can pin-

point what those effects were.

Many other successful missions were launched in the first two years of the ODP. The *Resolution* has traveled to the Pacific Ocean off the coast of South America and to the coast of northwest Africa, (where the ship's drills dug up an astounding 12,000 feet [3,660 meters] of the earth's core). On a mission in the mid-Atlantic, the researchers used video cameras like those used in the search for the *Titanic* to view a volcano jutting from the sea floor 2 miles (3.2 kilometers) down. On Leg 106, one of two

trips to the Mid-Atlantic Ridge, scientists were able to view hydrothermal vents complete with a large community of fascinating creatures.

Currently, the *Resolution* is in the midst of an 18-month jaunt in the Indian Ocean. "After that," says Rabinowitz, "we'll be exploring the Pacific for two years. Even with a ten-year mission, there's much more to explore than we'll have time for."

While scientists aboard the *Resolution* can study the world's long-term climatic changes through a

series of drilling missions planned well in advance and spanning several years, other ocean researchers studying certain short-term erratic, and devastating changes in the world's climate cannot preschedule their research at all. Those researchers are studying the unruly and sporadic climatic occurrence known as El Niño, in which the ocean and the atmosphere join, sometimes causing monsoons, droughts, and other disastrous events that can take place nearly anywhere on earth. All scientists know with certainty about the visi-

The JOIDES *Resolution* (left) and its crew are in the middle of a ten-year mission to study the earth's core through all the world's oceans. At right is the lab aboard the ship.

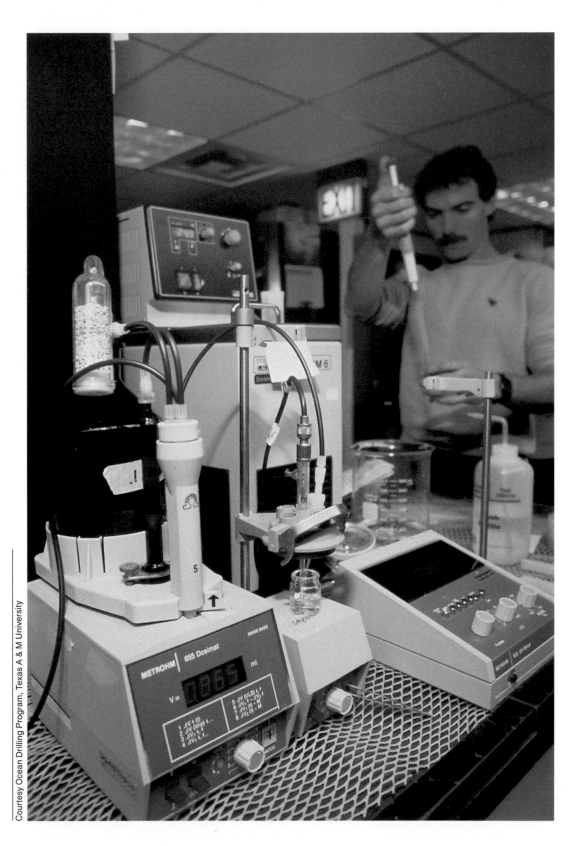

tations of El Niño is that it will surely recur—as it has at least nine times since World War II, most recently in 1986.

What exactly is El Niño, and what makes it so powerful? "Actually, until recently, it was thought to be a fascinating—but not terribly consequential—ocean-warming phenomenon," recalls Eugene Rasmusson, a faculty research associate at the University of Maryland's department of meteorology. "We knew that every few years, large areas of the Pacific ocean would undergo a dramatic, temporary

For many years, El Niño was considered a problem only for Peruvian fishermen (right). Now, we know that the ocean-weather phenomenon (left) can cause worldwide devastation.

© John Dominis/Wheeler Pictures

warming—but that was about all."

In fact, over the years El Niño has been of most interest to Peruvian anchovy fishermen and the manufacturers of guano, seabird droppings that are converted into fertilizer. It was the fishermen and merchants who named the phenomenon a century ago after the Christ child, with the Spanish words for "the Child," because of its usual appearance around Christmas.

For both the manufacturers and the fishermen, El Niño has been at best a nuisance and at worst an economic disaster. Their businesses depend on the cold, nutrient-rich water that usually wells up in the Pacific Ocean just off the coast of Peru. With this frigid current come swarms of anchovy, the fishermen's haul and the guano-producing cormorant birds' favorite food.

In years of El Niño, however, the temperature of the surface waters in crucial fishing areas becomes warm. The anchovies move away, leaving both the fishermen and the cormorants with nothing to catch. Millions of cormorants die, decimating the fertilizer industry, while anchovy boats return to port nearly empty-handed.

The effects of El Niño are not, however, isolated to the waters off the coast of Peru. El Niño also brings violently destructive rains—often accompanied by severe flooding in coastal areas that are usually desert dry, and the visitation has long been an integral part of life on the west coast of South America.

For El Niño is no new phenomenon. Researchers have tracked down evidence of conditions result-

W.H.O.I.

ing from El Niño as far back as 1726, working from sailors' and historians' journals that recount the mysterious and threatening weather patterns occurring at least once a decade and sometimes as frequently as every two or three years. Still, many of these visitations have been very mild—almost unnoticeable outside Peru—and it wasn't until El Niño's return in 1972 to 1973 that scientists began to treat it seriously.

In a bow to the past, El Niño's 1972 arrival brought on the near collapse of the Peruvian anchovy industry, which had grown enormously since the last severe occurrence in the 1950s. But this time the climatic phenomenon's reach was far greater than in the past. During this visit, El Niño brought vicious droughts to Africa, Australia, and elsewhere as well as caused devastating floods in the Philippines and other countries.

In fact, the weather nearly everywhere was so strange and destructive that year that it attracted the attention of more experts than ever before. "The damage caused by the 1972 to 1973 El Niño told us that we had to find out more about it," says Rasmusson. Thus, in the wake of this destruction the first in-depth study of El Niño began, using satellites able to judge air pressure and water temperature as well as computers and advanced meteorological devices.

What scientists found was an astounding ocean-atmosphere relationship that exists nearly worldwide. It is a relationship, says Rasmusson, that forces us to reevaluate our understanding of the role of the ocean in the long-term climate of the earth. "Until recently, most researchers saw El Niño as just a weather problem," he points out, "but now we know that the ocean plays a crucial—perhaps the most important—role in causing worldwide disruption of the weather."

In most years, winter weather patterns involve the presence of high-pressure weather systems stationed over the eastern Pacific Ocean off the coasts of the Ameri-

What do the ocean and violent storms around the world have in common? More than we ever imagined, assert scientists studying El Niño.

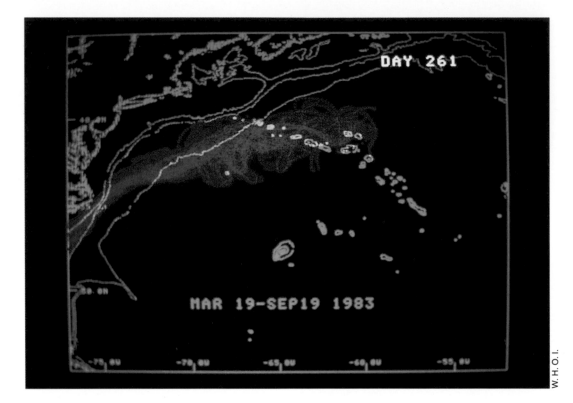

cas. Steady trade winds blow westward from these systems, actually pushing a current of warm water ahead of them toward Australia and other Southern Pacific lands. Once this warm surface water is pushed away, it is replaced by an upsurge of colder water—the nutrient-rich, frigid current so necessary to Peru's fishing industry. What follows is a year of comparatively predictable weather, as storms, high-pressure areas, and other systems build up over the ocean in response to the water's temperature.

That is the way the weather normally influences the ocean currents and is then influenced by it. In years of El Niño, however, everything gets turned upside down. The high-pressure area off South America weakens, while an area of low pressure in the western Pacific gains strength. The result: The westward trade winds disappear and are replaced by steady winds blowing east.

These winds, like the normal trade winds, then push the surface water ahead of them, creating an odd current from west to east—a current that only occurs in years of El Niño. This warm-water current flows toward South America in a phenomenon known as a Kelvin wave, disrupting the usual upwelling of cold water and actually raising the sea level off Peru.

A change in wind direction, an alteration of the current—none of it seems particularly threatening or important. But because such a large area is involved—perhaps a quarter of the earth—and weather is so closely tied to the temperature

Meteorologists use advanced instruments to carefully monitor air pressure and other phenomena that herald the arrival of El Niño (far left).

In 1982–83, disastrous floods (left) scoured normally dry areas in the United States, South America, and elsewhere—all the result of El Niño.

of oceans, the effects of these changes can be devastating. "Most El Niños are barely noticeable," says Rasmusson. "But every few years, one comes along that we can't ignore."

The most recent extremely destructive year was 1982, when El Niños wreaked havoc on five continents. Lasting through much of 1983, it must be ranked as one of the worst—and certainly the most widespread—natural disasters in history. "What made it even more disastrous was how unusual it was," Rasmusson points out. "We simply had no idea what would happen next."

In fact, for several months after these El Niño's began experts were still debating whether they had arrived at all. For example, higher than usual water temperatures

While some areas were drenched by endless rainstorms, other areas experienced intense droughts, accompanied by uncontrollable brush fires (left). When will the next destructive El Niño occur? No one knows.

were first noted in the western Pacific, instead of appearing off the Peruvian coast as usual. When the characteristic warm Kelvin wave arrived, however, this one raising the sea level a remarkable 8 inches (20 centimeters), scientists knew there was to be a massive visitation.

Within months, more than one thousand El Niño-related deaths (from storms or fire resulting from disastrous droughts) had been reported from South America, Australia, and Asia, and elsewhere, while tens of thousands of people were displaced or left homeless. The cost of property damage was astronomical, even without taking into account the aftermath effects felt for years in some areas.

Even more frightening than the latest El Niño's power was the extent and variety of its destruction. In Australia, it caused the most severe drought in this century, devastating the livestock industry and spurring enormous brushfires that often roared out of control for

weeks. Fierce droughts also afflicted Indonesia, Sri Lanka, and southern India and reached halfway across the world, to Mexico. Even normally moist Hawaii was not immune.

Drought was not the problem, however, in other areas. As in 1972 to 1973, Peru was again the center of seemingly endless rainstorms in normally dry areas. Ecuador, Bolivia, and parts of Brazil were also inundated in rain, suffering floods and mudslides as a result. Meanwhile, a series of extremely violent cyclones struck French Polynesia, and Hawaii—already suffering from drought—was the target of a sudden hurricane.

In the United States, El Niño's effects were not so severe. Still, the country's Gulf Coast was subjected to months of above-average rain, causing floods and enormous crop damage. On the West Coast, a winter of high snowfall and a spring of rain led to floods and mudslides in early 1983.

Eventually, normal ocean currents returned to the Pacific, and this El Niño—like all others—dissipated by mid-1983, leaving behind months of picking up the pieces. It also left scientists like Rasmusson wondering how to improve our ability to predict the coming of El Niño. "Luckily, we have trained oceanographers and meteorologists working together to design new prediction techniques."

These techniques, which involve complex computer models, got their first test in early 1986, when a new warming trend began. "This El Niño is very mild, but scientists saw it coming," Rasmusson says.

Even if researchers are someday able to predict the arrival of El Niño with perfect accuracy, they may never be able to predict its severity, Rasmusson warns. " We may be able to say that Australia may suffer from a drought," he admits. "But we'll never be able to tell individual farmers what that means for them."

FOOD FROM THE SEA

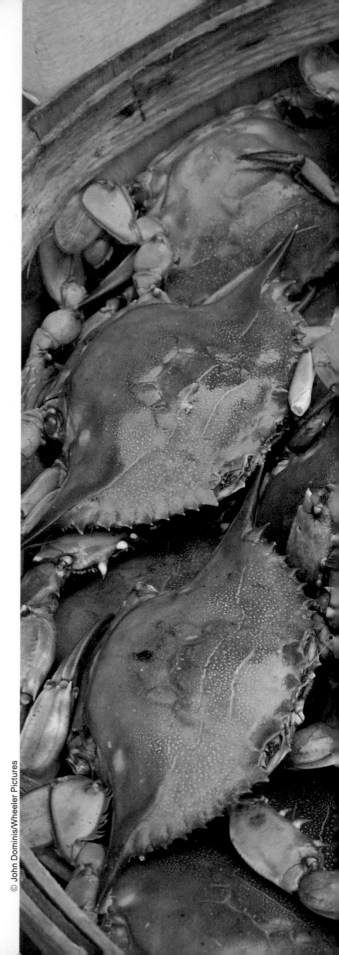

Long before *Alvin* first voyaged to the sea floor 2 miles (3.2 kilometers) below the surface, before the *Titanic* took off on its ill-fated maiden voyage, before treasure-laden Spanish galleons plied the waters of the Atlantic, before explorers took the first fragile wooden boats and set out to explore the vast unknown seas, people throughout the world depended on the ocean for food.

Coastal societies have always recognized the ocean as an invaluable provider. Until recently the harvest of the sea has been thought of as plentiful and far more dependable than that reaped from the land. Storms, cold spells, droughts, and any number of unexpected events can destroy months of hard work on a region's crops, while steady hunting can quickly ravish the surrounding lands of vital game.

The ocean, it was long thought, was different. All that was needed were stronger boats, new ways of tracking fish, and larger nets to gather an always abundant harvest; barely making a dent in what seemed to be an endless bounty. The oceans, in the eyes of earlier times, were so vast that there was apparently no way that they could ever be harmed by too much fishing.

That assumption of abundance may have been correct for thousands of years, but it didn't take the earth's startling population boom into account. More than five

Throughout history, the ocean has been an important and reliable source of food (page 115). Today, however, growing human populations are imposing new demands on its abundant offerings.

Destruction of coral reefs (left)—home to many of the world's fish—has led to declining fish populations and troubled times for fishermen (right).

Japan Information Center, Consulate General of Japan, N. Y.

billion people now inhabit the earth—more than twice as many as forty years ago. So its not surprising that in recent years we have seen many instances of a region's seemingly inexhaustible fisheries being destroyed by short-sighted overfishing to meet an escalating demand.

In Peru beginning in the 1950s—a decade already plagued by the visitation of an El Niño—fishermen added to the destructive impact of the phenomenon by harvesting so many anchovies that the population and the fishing industry has never recovered. Cormorants, with nothing to eat, nearly disappeared, which doomed the guano industry, so that Peru is now left with neither of its age-old sources of income. A recent decline in the numbers of fish-eating seabirds in the

Atlantic Ocean off the Canadian coast is a warning of similar ill health in that country's fish population. And many poor third-world countries, whose growing populations are far outstripping their ability to grow or buy food, are resorting to even more destructive techniques.

In the Philippines, for example, professional fishermen often choose an easy and illegal, though all too often unpoliced, way of harvesting their catch: dynamiting a coral reef. While tons of dead fish float obligingly to the surface, what's left behind is a mass of rubble that may never support life again.

While laws and subsequent enforcement can slow down these destructive activities, they cannot solve the basic problem: world hunger caused by a population that's

reached five billion and shows no sign of slacking off in growth. On land, any effective, long-term solution will require the development of hardier forms of grain, the teaching of more efficient farming techniques, and intensive conservation education to keep both farmers and major industries from destroying the land for short-term gain.

In the oceans, however, the challenges are equally great. Today, nearly a quarter of all animal protein eaten worldwide comes from fish, and in many countries ranging from Japan to many poor nations fish is the main food source. With catches remaining steady or declining, and demand growing—even in such beef-loving countries as the United States—the urge to cash in now and forget the future is hard to resist.

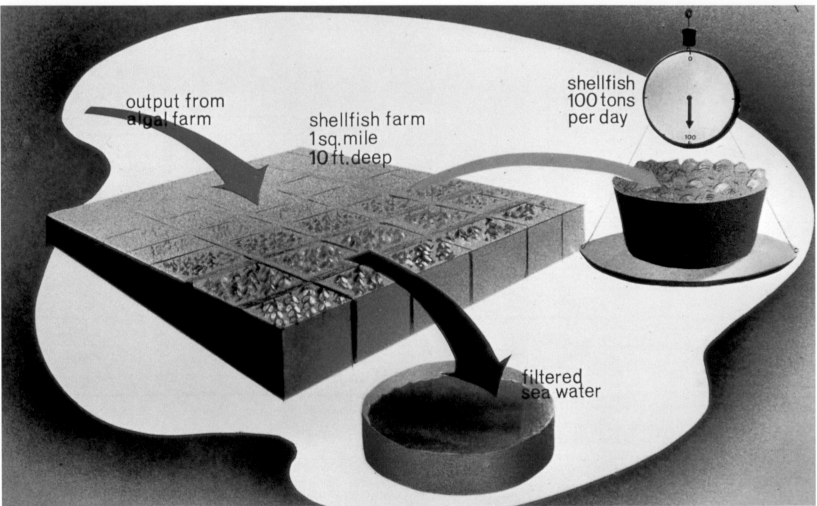

output from
algal farm

shellfish farm
1 sq. mile
10 ft. deep

shellfish
100 tons
per day

filtered
sea water

W.H.O.I.

Though hard hit, the oceans, lakes, rivers, and streams may still be able to provide the most exciting solution to the crises of increasing hunger, both today and in the future. That solution is fish farming, or *aquaculture*—and it is already one of the fastest-growing industries in the United States and many other countries.

Fish farming is not a new idea; in fact, it's been around for a remarkably long time. Most experts think the idea got its start four thousand years ago in China.

Those many years ago, the Chinese taste for fish, especially carp, quickly outweighed the natural supply. The ingenious Chinese solution to this shortage was to breed and grow the fish themselves.

Today, China continues to be one of the world's most successful aquacultural nations, harvesting more than three-quarters of a million tons of carp and other finfish (as opposed to shellfish) a year. Leading the pack is India, while Japan, the Soviet Union, Indonesia, and the Philippines follow. All

produce at least 100,000 pounds (45,300 kilograms) of farmed fish each year.

The list of the leading shellfish producers reads similarly, although European countries such as Spain, France and the Netherlands number among the top ten by growing vast quantities of mussels and oysters. China is the leader here, producing nearly 2 million tons (more than 1.5 million metric tons) of shellfish each year.

Africa and South America, continents whose fast-growing popula-

New research is allowing scientists to grow clams and other shellfish—and to provide more food to a hungry world (left). Here is an aquaculture pond (right).

W. H. O. I./John Ryther

tions would seem most in need of extensive fish-farming efforts, are just beginning to introduce programs; any major results are still years off. And the United States, even with all its wealth and resources, was for many years not much further ahead.

Today, due in part to money from the 1980 United States National Aquaculture Act and in part to a change in American food habits, the country has at last joined Europe and Asia in making a major effort to increase its yield from fish farming. And, internationally,

huge corporations, including British Petroleum, Campbell Soup, and Ralston-Purina, have entered the business, nearly guaranteeing that the amount of farmed fish and shellfish consumed worldwide—currently about 10 million tons (9 million metric tons), or roughly 15 percent of all fish consumed—will soon grow dramatically.

Of course, today's fish farmers don't just dig a hole, throw some fish in it, and hope that their "livestock" grows fat and stays healthy. The Division of Applied Biology at Harbor Branch Oceanographic In-

stitution in Florida, for example, has devoted a great deal of money and study to clam cultivation.

In this project—just one of thousands like it around the world—angel-wing clams are grown in outdoor pens, while scientists study growth rates, mortality, and ways of ensuring a productive harvest. Eventually, such research will help replenish supplies of many types of shellfish that have been driven to near extinction in the wild by both pollution and years of relentless harvesting.

Today, modern science is allow-

K.K. Chew, Div. of Fisheries, Science & Aquaculture/University of Washington

The oysters and other shellfish are cultivated in natural (left) or artificial (bottom right) beds.

ing some experts to produce "designer" fish and shellfish, animals alien to any wild stream or seabed. Through careful breeding and genetic engineering, these farmers are creating creatures that grow faster, taste better, and are easier to eat than their wild relatives.

Probably the most eccentric example of this genetic refinement is taking place on the grounds of the Clear Springs Trout Company in Idaho. Here, researchers are working to produce a boneless trout. Through careful breeding, the company has already cultivated a fish with far fewer bones than usual, as well as ones with a much higher proportion of flesh to bone. The living filet is bound to follow.

Researchers at the University of Washington in Seattle recently accomplished something nearly as dramatic. By manipulating oysters' chromosomes, they came up with a

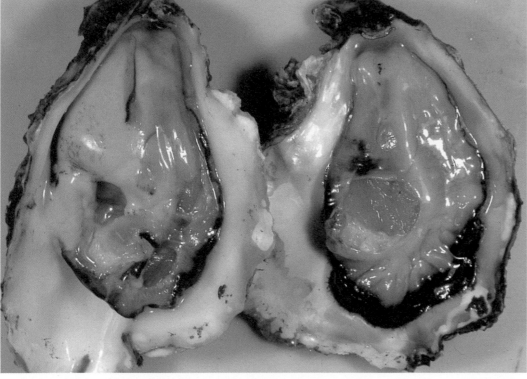

Stan Allen, Div. of Fisheries, Science & Aquaculture/University of Washington

The remarkable development of sterile oysters—which, unlike their fertile relatives, taste good year-round—is one of aquaculture's greatest recent advances. In the picture near left, the oyster on the right is sterile.

W. H. O. I.

sterile oyster. Why is this important? Because oysters become nearly inedible during their breeding season; a sterile one will taste good all year long.

Other aquacultural challenges are being met in less futuristic ways. Lobsters have long been difficult to cultivate because they have a bad habit: cannibalism. Put a crowd of young lobsters together in a tank, and eventually you'll be left with one fat lobster—and a failed lobster farm. Now, however, Maine fish-farmers have designed cages that keep the hungry juveniles apart, allowing all of them to grow to marketable size.

Surf clams, a popular commercial species, have also long been thought difficult to raise. When grown in artificial beds off the coast of the northeastern United States, they tended to be eaten by predators before reaching maturity.

The solution: fiberglass cages, complete with mesh that allows food- and oxygen-rich water to flow past, but keeps predators out.

While shrimp farming has perhaps become the most extensive venture of Central and South America into aquaculture, many tropical nations are far behind in the effort to develop fish farming. Particularly slow in this development have been the islands in the Caribbean Sea, many of which depend almost entirely on United States aid and tourism to support their shaky economies. Now, thanks to a research effort begun at the Smithsonian Institution in Washington, D.C., that may be changing.

During the past few years, scientists have been working to convert coral reef communities into suitable environments for fish farming. At first glance, reefs—overflowing with many fish species as well as lobsters, crabs, and other creatures—would seem a perfect place to attempt aquaculture. The problem lies in the fact that a reef is an extraordinarily complex system. Various fish, shellfish, and the coral itself depend on each other for food, in ways that would make it nearly impossible to remove and cultivate a single species.

Even more importantly, the coral reef and all its inhabitants depend on the constant flow of waves and currents to provide food and keep their environment healthy—and that is what gave the Smithsonian's Walter Adey a brilliant idea. Putting the open-ocean currents to use, he and others have designed plastic screens that sit just below

Courtesy Aquatic Systems, Inc., San Diego

the ocean surface in lagoons off the Bahamas and other Caribbean islands. Waves carry nutrients and green algae, a simple marine plant, past and through the screens, and the algae grabs hold of the screens and grows into a thick, luxuriant jungle.

Though the algae itself is nutritious, Adey and his coresearchers realized that the Caribbean and American diet would not soon welcome a slimy green plant to the dinner table. So they searched for a less picky eater for their algae— one that the cultured palate might in turn find tasty—and found the West Indian crab. Not only does this breed of crab thrive on the algae, it happens to taste delicious, as good, the researchers say, as the renowned Alaskan king crab.

With the worldwide natural harvest of lobsters (left) and shrimp (right) decreasing, fish farmers are struggling to find ways to breed the shellfish in captivity. Their major obstacle is that lobsters are cannibals and cannot be reared together in large enclosures.

Courtesy Aquatic Systems, Inc., San Diego

Fish farmers are currently studying every aspect of the lobster's life—from its microscopic beginnings (bottom right) to its adulthood (left)—in an effort to breed the animal more successfully.

Courtesy Aquatic Systems, Inc., San Diego

The ocean may one day be the world's major food source (right). But we must learn to farm this fragile environment more efficiently and to fish it more carefully.

The rest has been easy. First, the crabs' free-swimming larvae are kept for a few days in tanks. Then, when they're ready to settle down, they're placed in the open cages. Here they graze on the algae, growing to marketable size in about nine months. Caribbean fishermen then fetch a decent price for the crab meat.

In the future, other species that thrive on algae—including snails, fish, and sea urchins—will be farmed in the same way. According to Adey, this new method will be important for at least two reasons: It will provide an immediate source of work and income for island fishermen and will protect wild crabs and other creatures, many of which have been driven close to extinction by relentless overfishing.

In coming years, the worldwide fish-farming industry will surely continue to grow. There is no chance, however, that it will ever completely replace fishing as a means of bringing nutritious protein to the billions of people who depend on fish for food. So how will we manage to keep tuna, cod, and many other open-ocean fish—which are difficult, or impossible, to farm—from declining or disappearing entirely?

One answer currently gaining popularity is the artificial reef. This controversial idea is based on a simple concept: Where you find a reef, you always find more fish than you do in the open ocean. Therefore, build your own reefs where none exist—and then wait for the swarms of fish to make an appearance.

Although the first artificial reefs in the United States date back more than one hundred fifty years ago (fishermen sank structures made of logs to attract game fish in South Carolina), the possible economic benefit of these reefs was realized only gradually in the years that followed. And while fishing associations and local governments sometimes deliberately planted debris, many of the finest artificial reefs were created by accident.

For example, many of the hundreds of warships sunk during World War II are now teeming with sea life. In Truk Lagoon in the South Pacific, a famous harbor and battle site, dozens of wrecks have become magnets for swarms of colorful tropical fish. The metal stanchions of drilling rigs in the Gulf of Mexico also attract schools of large fish that might otherwise be scattered across open ocean.

Mark Atherton, courtesy Can-Dive Services Ltd.

© Jeff Simon/Wheeler Pictures

Undersea oil rigs (above) serve as artificial and unintentional reefs. Soon, they are populated by large numbers of food fish.

Japan Information Center, Consulate General of Japan, N. Y.

The Japanese are among the leaders in artificial reef and fish-farming technology, as this large-scale oyster farm demonstrates (left). One-fifth of Japan's huge coastline will soon harbor man-made reefs.

Sunken ships and other ocean debris (right) form artificial reefs that attract many fish. But do they actually cause an increase in fish populations? Scientists are not yet certain.

Even today, most of the artificial reefs off the United States look more like garbage dumps than carefully designed fish habitats. Rusty old cars, corroding tires, toilets—anything will serve as a home for fish. How good a home such materials provide—and how long these objects will survive underwater—is unknown.

As for the success of artificial reefs in the rest of the world, England, Australia, Israel and many other countries have just recently begun to experiment with designing reefs. The finest reefs of all, however, come from fish-loving Japan, where the first artificial reefs were built more than three hundred years ago. With the aid of the government, large expanses of reefs have been installed in recent years, and the ambitious Japanese master plan calls for the eventual placement of more than 2,500 new reefs along one-fifth of Japan's coastline.

Japanese reefs come in a variety of shapes and sizes that put most American models to shame. Some are towering concrete pyramids; others are complex terraces or cylinders. Each has been designed with a specific location and certain species of fish in mind. All seem to attract large numbers of the desired food fish.

However, neither America's rusting hulks nor Japan's concrete pyramids have yet provided the answer to an important question about artificial reefs: Do they really work? Of course, there's no question that these structures attract large numbers of tuna, mackerel, and other prized food fish. But are the reefs encouraging increases in fish populations or merely attracting and concentrating the existing fish populations so that they are caught more easily and therefore appear more plentiful?

Scientists simply don't know the answer to that question. If, in fact, artificial reefs act simply as magnets—and not as breeding grounds—then an idea first conceived to feed the world more successfully may only be hastening us toward an ocean barren of the food we've depended on for so long.

TOMORROW'S OCEAN

"We've learned so much about the deep sea in the past few years," says WHOI's Yoerger, "but what we've done is like drawing broad outlines on a map. What comes next—filling in the details—will be even more exciting."

For centuries, explorers sailed across the ocean's surface, risking their lives to discover new lands, but never looking beneath the surface to examine the world that thrived down below. Just fifty years ago, when the explorer Beebe descended in the first bathysphere, the sea and its creatures were almost a complete mystery to us—as distant and fascinating as the face of the moon. Even a decade ago, we knew more about life in the frigid wastes of the Arctic and the humid tangle of the rain forest than we did about the world that existed nearby our coastlines.

ANGUS, Alvin, Deep Rover, satellites, side-scan sonar, and dozens of other technological advances have at last enabled us to begin to comprehend the complexity and variety of life in the deep sea. WHOI's J. Frederick Grassle has shown us creatures that survive in a pitch-black world inundated with poisonous chemicals. Bruce Robison at the University of California has begun to probe the endless mysteries of the midwater. And Eugene Rasmusson has helped show why a late-spring snowfall in Washington, D.C., may be caused by changing water temperatures in the South Pacific.